G.I. Joe® Value Guide 1964—1978

Dolls, Gear & Equipment

by Carol Moody

Published by

Cumberland, Maryland 21502

Dedication

This book is dedicated to all the veterans who have defended our country. We admire and appreciate all they have sacrificed in the name of freedom. Although this book highlights the military of World War II, we hope to recognize all of the brave men and women who have served this great nation.

We thank you.

Acknowledgements

My deepest thanks to Lila Ayala for her detailed and thorough research assistance and photography, also to my husband, Gene, for his never-ending patience during this project and for his assistance in the photography. Special appreciation to C. Wayne Hildreth, who loaned his collection to be photographed.

Additional copies available at $12.95 each plus $1.75 postage from
Hobby House Press, Inc.
900 Frederick Street
Cumberland, MD 21502
or from your favorite bookstore or dealer.

Printed in the United States of America.
ISBN: 0-87588-330-3

Table of Contents

1. *G.I. Joe Action Soldier.*

2. *G.I. Joe Action Soldier.*

I. G.I. Joe — The Beginning

G.I. Joe was popular — yet controversial during 1964-1975 (the Vietnam era). As the war went on, people began to boycot *G.I. Joe* feeling that he represented violence. Therefore, the military series was discontinued and the *"Adventure Team"* began.

Joe has gone through a number of changes. He began as a World War II fighting man whose face was comprised of the faces of 20 Medal of Honor recipients. From there he entered the Space Age and was an adventurer.

Joe has also gone through several physical changes. He started as a 12in (31cm) fully-jointed fighting man, was reduced to an 8in (20cm) *Super Joe*, and further reduced to the present size of 3in (8cm) and is called a "Mobile Strike Force Team."

II. G.I. Joe — 1964-1976

G.I. Joe, a product of Hassenfeld Bros., Inc., was introduced in 1964 and produced through 1976 as a 12in (31cm) plastic figure.

The first *G.I. Joe* had painted hair, either blonde, brown, black or red. His eyes were blue or brown. The hair and eye colors came in all combinations; no discrimination was made as to which branch of the service had which hair or eye color.

The faces have a scar on the right cheek, except the "Action Soldiers of the World." These do not have scars and the heads are smaller and a different mold was used. There were four branches of the service represented: Army, Navy, Air Force and Marines.

The year 1965 saw two new innovations for *G.I. Joe*; a black *Joe* and the first vehicle scaled to fit him — a Jeep.

In 1966 the *G.I. Joe* "Action Soldiers of the World" were introduced. The figures in this series were: *German Soldier, Japanese Imperial Soldier, Russian Infantry Man, French Resistance Fighter, British Commando* and *Australian Jungle Fighter*. Also in this year the emphasis was on equipment for *Joe*, covering a much broader scale. The figures could be purchased alone or in sets with equipment and accessories. In this year *Joe* joins the space program and the Vietnam War, with a Space Capsule and a Green Beret set.

1967 brought the introduction of the talking *Joe*. *G.I. Nurse* was made this year only and this was the last year for the "Combat Series."

In 1968 nothing new was produced but the "Action Soldiers of the World" series was discontinued.

The year 1969 begins to lean away from military influence and toward adventure with the introduction of the *G.I. Joe* "Adventure Series," including the *Talking Astronaut*.

Beginning in 1970 the focus for *Joe* was on adventure with the introduction of the "Adventure Team Series," including *Adventure Team Commander, Black Adventurer* and *Adventure Set Land Adventurer.* This year also saw the introduction of the flocked hair and beard for *Joe* in blonde, brunette and as a redhead.

During 1971 *Joe* had every kind of adventure imaginable, from Smoke Jumper to Karaté, with this trend continuing through 1977.

In 1974 talking *G.I. Joe* had flocked hair and Kung Fu Grip hands.

With 1975 we see a much different type of figure from Hasbro Industries, Inc. This figure has the same body as *Joe* but a different head. He is called *Atomicman.* He has Kung Fu Grip hands. His left leg and right arm are clear robot-like limbs. He has brown hair and a small hole in the top of his head that you can look into and see out of his right eye.

The *Intruder* was introduced in 1976. It had an ape-like body, white eyes and a button to push in his back to move his arms in a "Crusher Grip." Also in 1976 came the *Eagle Eye Joe* with movable eyes. The eyes moved by means of a lever on the back of the head. He had molded-on blue trunks and the regular *G.I. Joe* head mold with fuzzy hair and beard.

The *Defender* was also introduced during the busy year of 1976. The head mold was different and the arms and legs were jointed at the hip and shoulder only. He no longer said *G.I. Joe* on his buttocks.

In 1978 domestic marketing of *G.I. Joe* was discontinued.

3. *G.I. Joe Action Marine.*

4. *Japanese Imperial Soldier* from the "Action Soldiers of the World" series.

III. Super Joe — 1977

Beginning in 1977 Hasbro Industries, Inc., came out with a new *Joe*. He was 8in (20cm), fully jointed with hands that grip, painted hair and beard and blue molded trunks. Two push buttons on his back operated his arms.

From 1982 to the present we see the third phase of *G.I. Joe* in the form of a 3in (8cm) action figure. The focus here is on teamwork in the form of a "Mobile Strike Force" to combat evil in the world. They perform as a group instead of individuals; however, the theme has come full circle - back to fighting soldiers.

IV. Illustration — Figures only, dates, body marks, stock numbers

1964, 1965, 1966 — Body marks on buttocks
G.I. Joe™
Copyright 1964
By Hasbro®
Patent Pending
Made in U.S.A.
All figures listed below have these markings and painted hair:

#7500 — *Action Soldier* dressed in green fatigues, black boots, green cap, dog tag, insignia set, with training manual.

#7600 — *Action Sailor* dressed in blue work shirt and pants, black boots, white cap, dog tag, insignia set, with training manual.

#7700 — *Action Marine* dressed in camouflaged fatigues, black boots, green cap, dog tag, insignia set, with training manual.

#7800 — *Action Pilot* dressed in orange flight suit, black boots, dog tag, insignia set, with training manual.

#7900 — Black *G.I. Joe*, dressed in green fatigues, black boots, green cap, dog tag, insignia set, with training manual.

1964, 1965, 1966, 1967 — Body marks on buttocks
G.I. Joe®
Copyright 1964
By Hasbro®
Patent Pending
Made in U.S.A.
All figures listed below have these markings and painted hair:
#7500, 7600, 7700, 7800, 7900, dressed same as above listed figures.

#8100 — *German Soldier* dressed in dark green flannel jacket and pants, black calf-length boots, gray helmet, Luger pistol, holster, cartridge belt, woolly field pack, 9mm Schmeisser machine gun, six hand grenades, Iron Cross medal, manual. 1966-1967.

#8101 — *Japanese Imperial Soldier* dressed in khaki jacket and pants, helmet, brown boots, Nambu pistol, holster, Arisaka rifle with bayonet, field pack, cartridge belt, Order of the Kite medal, manual. 1966-1967.

#8102 — *Russian Infantry Man* dressed in dark green jacket and pants, fur hat, black calf-length boots, machine gun with bi-pod, field glasses and case, anti-tank grenades, ammo box, Order of Lenin medal, manual. 1966-1967.

#8103 — *French Resistance Fighter* dressed in black knit sweater, black pants, black boots, black beret, Lebel revolver, shoulder holster, radio set with earphones, 7.65 MAS machine gun, knife, grenades, Croix de Guerre medal, manual. 1966-1967.

#8104 — *British Commando* dressed in dark green flannel jacket and pants, brown boots, helmet, gas mask, Sten Mark 2S sub-machine gun and case, canteen and cover, Victoria Cross medal, manual. 1966-1967.

#8105 — *Australian Jungle Fighter* dressed in khaki bush jacket and shorts, knee-length olive green socks and brown boots, grenades, flamethrower, jungle knife, machete and sheath, entrenching tool, Victoria Cross medal, manual. 1966-1967.

5. *German Soldier* from the "Action Soldiers of the World" series.

6. *G.I. Joe Man of Action* with black "life-like" hair.

7. *German Soldier.*

1967, 1968, 1969, 1970 — Body marks on buttocks
G.I. Joe®
Copyright 1964
By Hasbro®
Pat. No. 3,277,602
Made in U.S.A.
All figures listed below have these markings:

#1915 — *Talking Astronaut* dressed in white coveralls, white boots, dog tag, insignia, manual. 1969-1970.

#7400 — *Talking Adventure Team Commander* with flocked hair and beard, dressed in green jacket and pants, black boots, pistol, shoulder holster, insignia. 1970.

#7401 — *Land Adventurer* with flocked hair and beard, dressed in camouflage fatigues, black boots, pistol, shoulder holster, dog tag, insignia. 1970.

#7404 — *Black Adventurer* with flocked hair, no beard, dressed in tan shirt and pants, black boots, pistol, shoulder holster, dog tag, insignia. 1970.

#7500 — *G.I. Joe Man of Action* with flocked hair, no beard, dressed in green army fatigues, hat, black boots, dog tag, insignia. 1970.

#7590 — *Talking Soldier* with painted hair, dressed in green fatigues, green cap, black boots, dog tag, insignia, manual. 1967.

#7690 — *Talking Sailor* with painted hair, dressed in blue work shirt, blue pants, white cap, black boots, dog tag, insignia, manual. 1967.

#7790 — *Talking Marine* with painted hair, dressed in camouflaged fatigues, green cap, black boots, insignia, manual. 1967.

#7890 — *Talking Pilot* with painted hair, dressed in orange zip-front flight suit, blue cap, black boots, dog tag, insignia, manual. 1967.

#7905 — *Adventurer* with flocked hair, no beard, dressed in blue denim shirt and pants, black boots, pistol, shoulder holster, dog tag. 1970.

1967 Only — Body marks on lower back
Patent Pending
©1967 Hasbro
Made in Hong Kong
#8060 — *G.I. Nurse* with blonde rooted hair and eyelashes, green eyes, dressed in white nurse's uniform, white cap, white shoes, medical bag, stethoscope, plasma bottle, two crutches, package of bandages, splints. 1967.

1970-1974 — Body marks on back
G.I. Joe Reg.™
© R.D. 1964
Hasenfeld Br. Inc.
Pat. 1966
#7292 — *Talking G.I. Joe* with Kung Fu Grip, flocked hair, no beard, dressed in green fatigues, black boots, dog tag, insignia, rifle, manual. 1974.

1974-1977 — Body marks
©1974
Hasbro Ind Inc.
Pawtucket. R. I. 02861
Made in Hong Kong (on buttocks)
Defender with hollow plastic rotationally molded body, jointed at hips, shoulders and head only. No molded-on trunks. Same head mold as *Bulletman.* Camouflage shorts only, tagged " 'Defender'™ by Hasbro, Hong Kong." No shoes. 1975.

1975-1977 — Body marks
©1975 Hasbro
Pat. Pend. Pawt. R.I. (on back of torso)
Made in Hong Kong (on back of neck)
Eagle Eye with flocked hair and beard, lever on back of head moves eyes from side to side, molded blue shorts. 1975-1977.

8. *Talking Action Sailor.*

9. *Action Marine.*

10. Black *Adventurer* with hair.

11. *Action Marine.*

#7270 — *Adventure Team*, with flocked hair and beard, molded blue shorts. 1975-1977.
#8026 — *Bulletman* with both arms and hands silver, different head mold, molded blue shorts, red sleeveless bodysuit, red boots, silver bullet helmet. 1976.

1975-1977 — Body marks
G.I. Joe®
Copyright 1964
by Hasbro®
Pat. No. 3,277,602
Made in U.S.A. (on buttocks)
Made in Hong Kong (on back of neck)
#8025 — *Atomicman* (Mike Power) with the same body as the painted hair and flocked hair figures, different head mold, Kung Fu Grip hands, left leg and right arm are clear robot-like limbs, see-through eye, dressed in camouflaged shirt, brown shorts.
Shirt is tagged "Atomicman." 1975.

1976 — Body marks
©1976 Hasbro®
Pat. Pend (on back of torso)
Hong Kong (on back of neck)
Intruder with squatty gorilla-type body, white eyes, button on back to make his arms grip. There were two different *Intruders*. Both wore an armor-styled bodysuit. The *Intruder* with the gold bodysuit had a beard; the one with the silver bodysuit had no beard. 1976.

V. Illustration — Gear and Equipment, stock numbers.

#7501 — Combat Field Jacket Set
#7502 — Combat Field Pack Set
#7503 — Combat Fatigue Shirt
#7504 — Combat Fatigue Pants
#7505 — Combat Field Jacket

#7506 — Combat Field Pack
#7507 — Combat Helmet Set
#7508 — Combat Sand Bag Set
#7509 — Combat Mess Kit Set
#7510 — Combat Rifle Set
#7511 — Combat Camouflage Netting Set
#7512 — Bivouac, Sleeping Bag Set
#7513 — Deluxe Pup Tent Set
#7514 — Bivouac Machine Gun Set
#7515 — Bivouac Sleeping Bag
#7516 — Sabotage Set
#7517 — Command Post Poncho Set
#7518 — Command Post Small Arms Set
#7519 — Command Post Poncho
#7520 — Command Post Field Set
#7521 — Military Police Set
#7522 — Jungle Fighter Card
#7523 — M. P. Duffle Bag
#7524 — M. P. Ike Jacket Set
#7525 — M. P. Ike Pants
#7526 — M. P. Helmet and Small Arms Set
#7527 — Ski Patrol
#7528
#7529
#7530 — Mountain Troops
#7531 — Ski Patrol Set
#7532 — Special Forces Bazooka Set
#7533 — Green Beret Card
#7534
#7535
#7536 — Green Beret and Equipment
#7537 — West Point Cadet Set
#7538 — Heavy Weapons Set
#7601 — Sea Rescue Set
#7602 — Frogman Set
#7603 — Frogman Scuba Suit Top
#7604 — Frogman Scuba Suit Bottom
#7605 — Frogman Equipment
#7606 — Frogman Scuba Tank

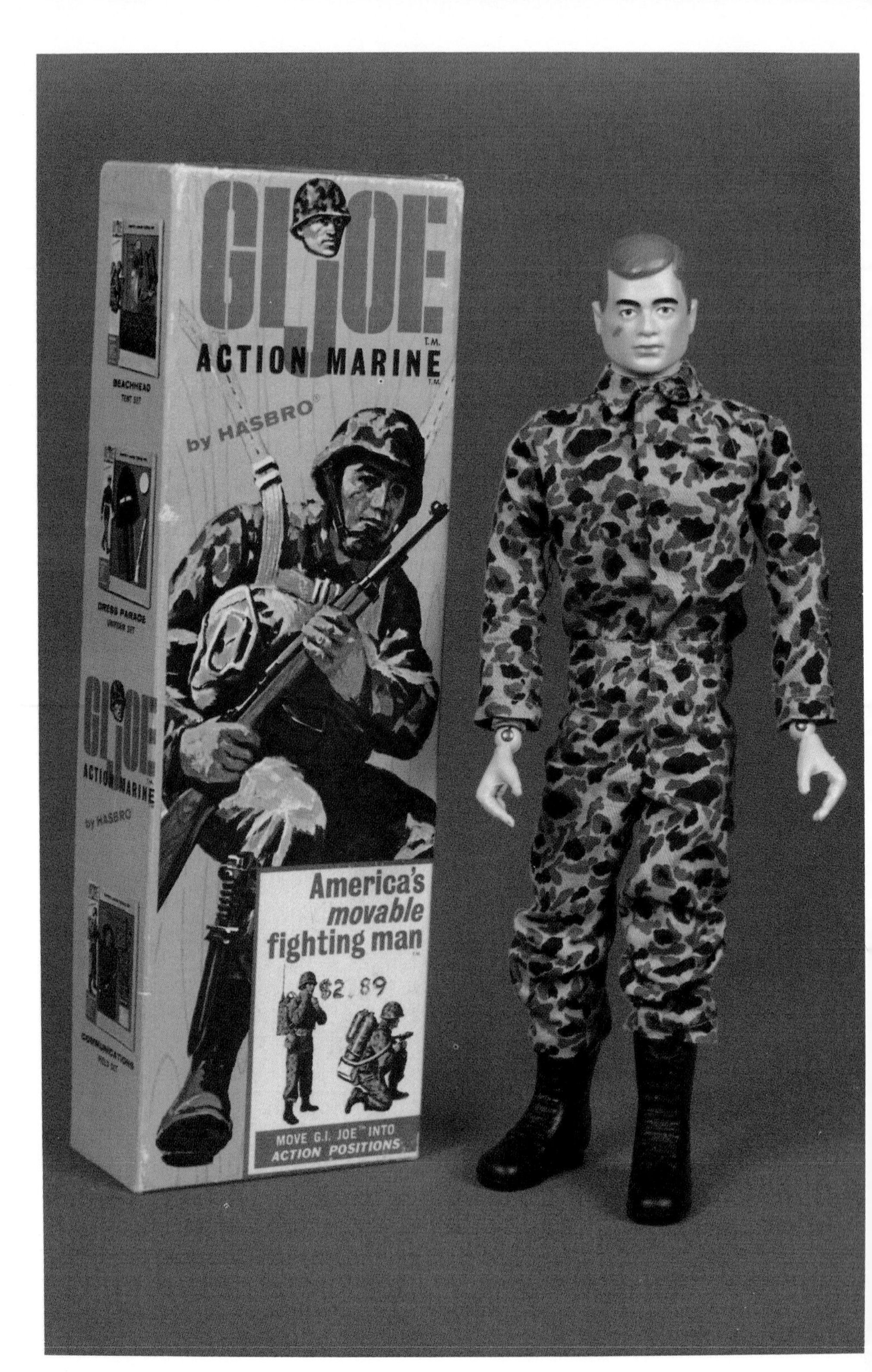

12. *Action Marine.*

13. *Action Soldier.*

#7607 — Navy Attack Set
#7608 — Navy Attack Work Shirt
#7609 — Navy Attack Dungaree Pants
#7610 — Navy Attack Helmet Set
#7611 — Navy Attack Life Jacket
#7612 — Shore Patrol Set
#7613 — Shore Patrol Jumper
#7614 — Shore Patrol Pants
#7615 — Shore Patrol Seabag
#7616 — Shore Patrol Small Arms
#7617
#7618 — Machine Gun
#7619 — Dress Parade
#7620 — Deep Sea Diver
#7621 — Landing Signal Officer Set
#7622 — Sea Rescue Deck Commander
#7623 — Deep Freeze Set
#7624 — Annapolis Cadet
#7625 — Breeches Buoy
#7626 — LSO Card
#7627 — Life Ring Card
#7701 — Communications Post Set
#7702 — Communications Poncho
#7703 — Communications Field Set
#7704 — Communications Flag Set
#7705 — Paratrooper Parachute Pack Set
#7706 — Paratrooper Small Arms Set
#7707 — Paratrooper Helmet Set
#7708 — Paratrooper Camouflage Set
#7709 — Paratrooper Parachute Pack
#7710 — Marine Dress Parade Set
#7711 — Beachhead Assault Tent Set
#7712 — Beachhead Assault Field Pack Set
#7713 — Beachhead Assault Field Pack
#7714 — Beachhead Assault Fatigue Shirt
#7715 — Beachhead Assault Fatigue Pants
#7716 — Beachhead Assault Mess Kit
#7717 — Beachhead Assault Rifle Set
#7718 — Beachhead Assault Flamethrower

#7719 — Marine Medic Set
#7720 — Medic
#7721 — First Aid
#7722
#7723 — Bunk Bed Card
#7724
#7725 — Heavy Weapons Set (same as #7538)
#7726
#7727 — Weapons Rack
#7728
#7729
#7730 — Demolition Set
#7731 — Tank Commander
#7732 — Jungle Fighter
#7801 — Survival Equipment Set
#7802 — Survival Life Raft
#7803 — Air Force Dress Uniform Set
#7804 — Dress Jacket
#7805 — Dress Pants
#7806 — Air Force Shirt
#7807 — Scramble Set
#7808 — Scramble Flight Suit
#7809 — Scramble Life Vest
#7810 — Scramble Helmet
#7811 — Scramble Parachute Pack
#7812 — Communications Set
#7813 — Air Police
#7814
#7815 — Air Security
#7816
#7817
#7818
#7819
#7820 — Fire Fighter or Crash Crew Set
#7821
#7822 — Air Cadet Set
#7823 — Fighter Pilot
#7824 — Astronaut Suit
#7825 — Air/Sea Rescue

14. Black *Adventurer* with life-like hair.

15. *G.I. Joe Man of Action.*

G.I. Joe® Price Guide

by Carol Moody

The prices listed here are for "Never Removed From Box" figures and gear and figures and gear in "Excellent condition, without box." No prices are given for dirty or damaged figures, clothing and accessories.

These prices are intended as a "guide" only. Please allow for availability and popularity in your particular collecting area. (E.C. is in original clothing.)

1989 G.I. Joe® Figure Price Guide

(Stock numbers without values are not listed.)

Stock Number	Name of Figure	N.R.F.B. Value	E.C. Value
1915	*Talking Astronaut* (1969-1970)	$200.00	$100.00
7270	*Adventure Team* (1975-1977)	150.00	75.00
7292	*Talking G.I. Joe* with Kung-Fu Grip (1974)	175.00	90.00
7400	*Talking Adventure Team Commander* (1970)	175.00	90.00
7401	*Land Adventurer* (1970)	150.00	75.00
7404	*Black Adventurer* (1970)	225.00	125.00
7500	*G.I. Joe Man of Action* (1970)	140.00	75.00
7500	*Action Soldier* (1964)	225.00	125.00
7590	*Talking Soldier* (1967)	200.00	125.00
7600	*Action Sailor* (1964)	235.00	135.00
7690	*Talking Sailor* (1967)	210.00	145.00
7700	*Action Marine* (1964)	225.00	125.00
7790	*Talking Marine* (1967)	200.00	125.00
7800	*Action Pilot* (1964)	235.00	135.00
7890	*Talking Pilot* (1967)	200.00	100.00
7900	*Black Action Soldier* (1964)	750.00	375.00
7905	*Adventurer* (1970)	150.00	75.00
8025	*Atomicman (Mike Power)* (1975)	80.00	35.00
8026	*Bulletman* (1977)	100.00	45.00
8060	*G.I. Nurse* (1967)	750.00	375.00
8100	*German Soldier* (1966-1967)	250.00	150.00
8101	*Japanese Imperial Soldier* (1966-1967)	275.00	175.00
8102	*Russian Infantry Man* (1966-1967)	250.00	150.00
8103	*French Resistance Fighter* (1966-1967)	250.00	150.00
8104	*British Commando* (1966-1967)	250.00	150.00
8105	*Australian Jungle Fighter* (1966-1967)	250.00	150.00
----	*Intruder* (1976)	80.00	35.00
----	*Super Joe*	50.00	25.00
----	*Eagle Eye* (1975-1977)	75.00	35.00
----	*Defender* (1974)	30.00	15.00

Stock Number	Name of Gear or Equipment	N.R.F.B. Value	E.C. Value
7501	Combat Field Jacket Set	$35.00	$15.00
7502	Combat Field Pack Set	35.00	15.00
7503	Combat Fatigue Shirt	25.00	10.00
7504	Combat Fatigue Pants	20.00	8.00
7505	Combat Field Jacket	25.00	10.00
7506	Combat Field Pack	20.00	8.00
7507	Combat Helmet Set	20.00	8.00
7508	Combat Sand Bag Set	20.00	9.00
7509	Combat Mess Kit Set	25.00	10.00

7510	Combat Rifle Set	25.00	10.00
7511	Combat Camouflage Netting Set	25.00	12.00
7512	Bivouac, Sleeping Bag Set	30.00	15.00
7513	Deluxe Pup Tent Set	35.00	15.00
7514	Bivouac Machine Gun Set	40.00	20.00
7515	Bivouac Sleeping Bag	20.00	8.00
7516	Sabotage Set	40.00	25.00
7517	Command Post Poncho Set	35.00	15.00
7518	Command Post Small Arms Set	35.00	15.00
7519	Command Post Poncho	25.00	12.00
7520	Command Post Field Set	30.00	10.00
7521	Military Police Set	60.00	40.00
7522	Jungle Fighter Card	25.00	12.00
7523	M. P. Duffle Bag	20.00	10.00
7524	M. P. Ike Jacket Set	35.00	15.00
7525	M. P. Ike Pants	25.00	10.00
7526	M. P. Helmet and Small Arms Set	35.00	15.00
7527	Ski Patrol	45.00	25.00
7530	Mountain Troops	45.00	25.00
7531	Ski Patrol Set	60.00	40.00
7532	Special Forces Bazooka Set	45.00	30.00
7533	Green Beret Card	30.00	15.00
7536	Green Beret and Equipment	60.00	40.00
7537	West Point Cadet Set	175.00	100.00
7538	Heavy Weapons Set	45.00	25.00
7601	Sea Rescue Set	35.00	20.00
7602	Frogman Set	40.00	25.00
7603	Frogman Scuba Suit Top	30.00	15.00
7604	Frogman Scuba Suit Bottom	25.00	12.00
7605	Frogman Equipment	25.00	12.00
7606	Frogman Scuba Tank	20.00	10.00
7607	Navy Attack Set	40.00	25.00
7608	Navy Attack Work Shirt	20.00	8.00
7609	Navy Attack Dungaree Pants	20.00	8.00
7610	Navy Attack Helmet Set	25.00	10.00
7611	Navy Attack Life Jacket	20.00	10.00
7612	Shore Patrol Set	90.00	45.00
7613	Shore Patrol Jumper	40.00	20.00
7614	Shore Patrol Pants	30.00	15.00
7615	Shore Patrol Seabag	20.00	8.00
7616	Shore Patrol Small Arms	35.00	20.00
7618	Machine Gun	40.00	20.00
7619	Dress Parade	60.00	30.00
7620	Deep Sea Diver	40.00	25.00
7621	Landing Signal Officer Set	45.00	25.00
7622	Sea Rescue Deck Commander	40.00	20.00
7623	Deep Freeze Set	50.00	35.00
7624	Annapolis Cadet	185.00	110.00
7625	Breeches Buoy	20.00	8.00
7626	LSO Card	20.00	8.00
7627	Life Ring Card	20.00	8.00
7701	Communications Post Set	25.00	12.00
7702	Communications Poncho	25.00	12.00
7703	Communications Field Set	25.00	12.00
7704	Communications Flag Set	30.00	20.00
7705	Paratrooper Parachute Pack Set	40.00	20.00
7706	Paratrooper Small Arms Set	35.00	15.00

16. *Talking G.I. Joe Man of Action* with life-like hair.

17. *Talking G.I. Joe Adventure Team Commander* with life-like hair and beard.

7707	Paratrooper Helmet Set	20.00	10.00
7708	Paratrooper Camouflage Set	25.00	12.00
7709	Paratrooper Parachute Pack	35.00	15.00
7710	Marine Dress Parade Set	125.00	90.00
7711	Beachhead Assault Tent Set	30.00	15.00
7712	Beachhead Assault Field Pack Set	35.00	15.00
7713	Beachhead Assault Field Pack	30.00	15.00
7714	Beachhead Assault Fatigue Shirt	20.00	10.00
7715	Beachhead Assault Fatigue Pants	20.00	8.00
7716	Beachhead Assault Mess Kit	20.00	10.00
7717	Beachhead Assault Rifle Set	25.00	12.00
7718	Beachhead Assault Flamethrower	30.00	15.00
7719	Marine Medic Set	45.00	25.00
7720	Medic	30.00	15.00
7721	First Aid	30.00	15.00
7723	Bunk Bed Card	20.00	10.00
7725	Heavy Weapons Set (same as #7538)	45.00	25.00
7727	Weapons Rack	40.00	20.00
7730	Demolition Set	35.00	15.00
7731	Tank Commander	50.00	25.00
7732	Jungle Fighter	40.00	20.00
7801	Survival Equipment Set	35.00	15.00
7802	Survival Life Raft	20.00	10.00
7803	Air Force Dress Uniform Set	125.00	90.00
7804	Dress Jacket	45.00	25.00
7805	Dress Pants	35.00	20.00
7806	Air Force Shirt	30.00	10.00
7807	Scramble Set	65.00	40.00
7808	Scramble Flight Suit	25.00	12.00
7809	Scramble Life Vest	20.00	8.00
7810	Scramble Helmet	20.00	10.00
7811	Scramble Parachute Pack	45.00	25.00
7812	Communications Set	25.00	12.00
7813	Air Police	50.00	30.00
7815	Air Security	30.00	15.00
7820	Fire Fighter or Crash Crew Set	50.00	30.00
7822	Air Cadet Set	125.00	90.00
7823	Fighter Pilot	45.00	25.00
7824	Astronaut Suit	50.00	40.00
7825	Air/Sea Rescue	35.00	20.00

Stock Number	**Name of Gear or Equipment**	**N.R.F.B. Value**	**E.C. Value**
7000	Five Star Army Jeep with Trailer	$175.00	$85.00
8000	Foot Locker	30.00	15.00
8020	Astronaut Suit and Space Capsule Set	250.00	150.00
8030	Desert Patrol Jeep	200.00	125.00
8040	Crash Crew Fire Set	400.00	200.00
8050	Frogman Sea Sled Set	90.00	45.00
*	Military Staff Car	350.00	200.00
*	Motorcycle	200.00	100.00
*	Side Car/Cycle	250.00	125.00
*	Personnel Carrier/Mine Sweeper	250.00	125.00
*	Navy Jet	375.00	200.00
*	Carry Case/Sentry Post (Sears Exclusive)	90.00	45.00

* - Stock numbers unknown.

18. Stock number 7500 *Action Soldier*, painted black hair, blue eyes, dressed in green fatigues, black boots, green cap, dog tag, with insignia set, Army manual. 1964.

19. Stock number 7500 *Action Soldier,* painted blonde hair, brown eyes, dressed in green fatigues, black boots, green cap, dog tag, with insignia set, Army manual. 1964. *Lila Ayala Collection.*

20. Stock number 7600 *Action Sailor,* painted black hair, blue eyes, dressed in blue work shirt and pants, black boots, white cap, dog tag, insignia set, Navy manual. 1964. *Lila Ayala Collection.*

21. Stock numbers 7613 and 7614, Shore Patrol Jumper with kerchief, Shore Patrol pants. *Lila Ayala Collection.*

22. *Talking G.I. Joe Action Sailor.*

23. Combat Gear and Equipment.

24. Stock number 7700 *Action Marine,* painted red hair, blue eyes, dressed in camouflaged fatigues, green cap, black boots, dog tag, with insignia set, Marine manual. 1964.

25. Stock number 7700 *Action Marine,* painted black hair, brown eyes, dressed in camouflaged fatigues, green cap, black boots, dog tag, with insignia set, Marine manual. 1964. *Lila Ayala Collection.*

26. Stock number 7900 *Black Action Soldier,* painted black hair, brown eyes, dressed in green fatigues, brown boots, green cap, dog tag, with insignia set, Army manual. 1964. *Eva Thompson Collection.*

27. Stock number 7800 *Action Pilot,* painted brown hair, dressed in orange flight suit, blue cap, black boots, dog tag, with insignia set, Air Force manual. 1964. Illustration from color brochure by *Hassenfeld Bros., Inc.*

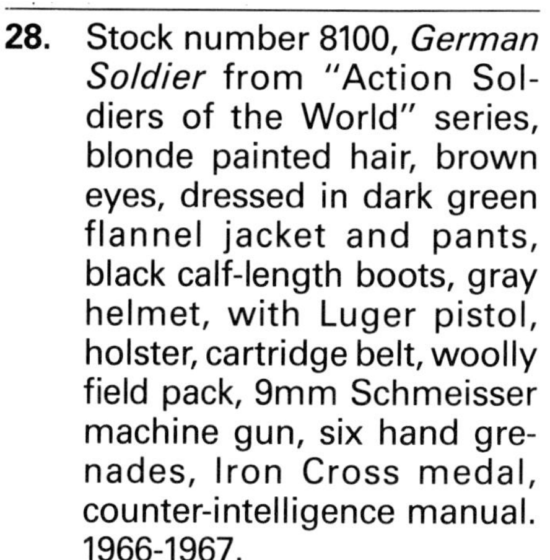

28. Stock number 8100, *German Soldier* from "Action Soldiers of the World" series, blonde painted hair, brown eyes, dressed in dark green flannel jacket and pants, black calf-length boots, gray helmet, with Luger pistol, holster, cartridge belt, woolly field pack, 9mm Schmeisser machine gun, six hand grenades, Iron Cross medal, counter-intelligence manual. 1966-1967.

29. Stock number 8101, *Japanese Imperial Soldier* from "Action Soldiers of the World" series, black painted hair, brown eyes, dressed in khaki jacket and pants, green helmet, brown boots, with Nambu pistol, holster Arisaka rifle with bayonet, field pack, cartridge belt, Order of the Kite medal, counter-intelligence manual. 1966-1967.

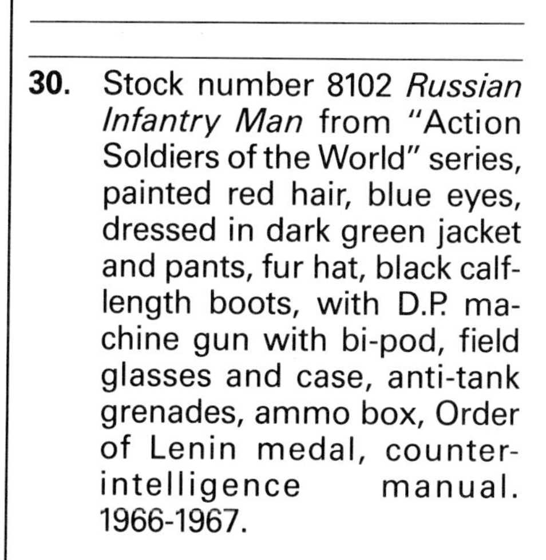

30. Stock number 8102 *Russian Infantry Man* from "Action Soldiers of the World" series, painted red hair, blue eyes, dressed in dark green jacket and pants, fur hat, black calf-length boots, with D.P. machine gun with bi-pod, field glasses and case, anti-tank grenades, ammo box, Order of Lenin medal, counter-intelligence manual. 1966-1967.

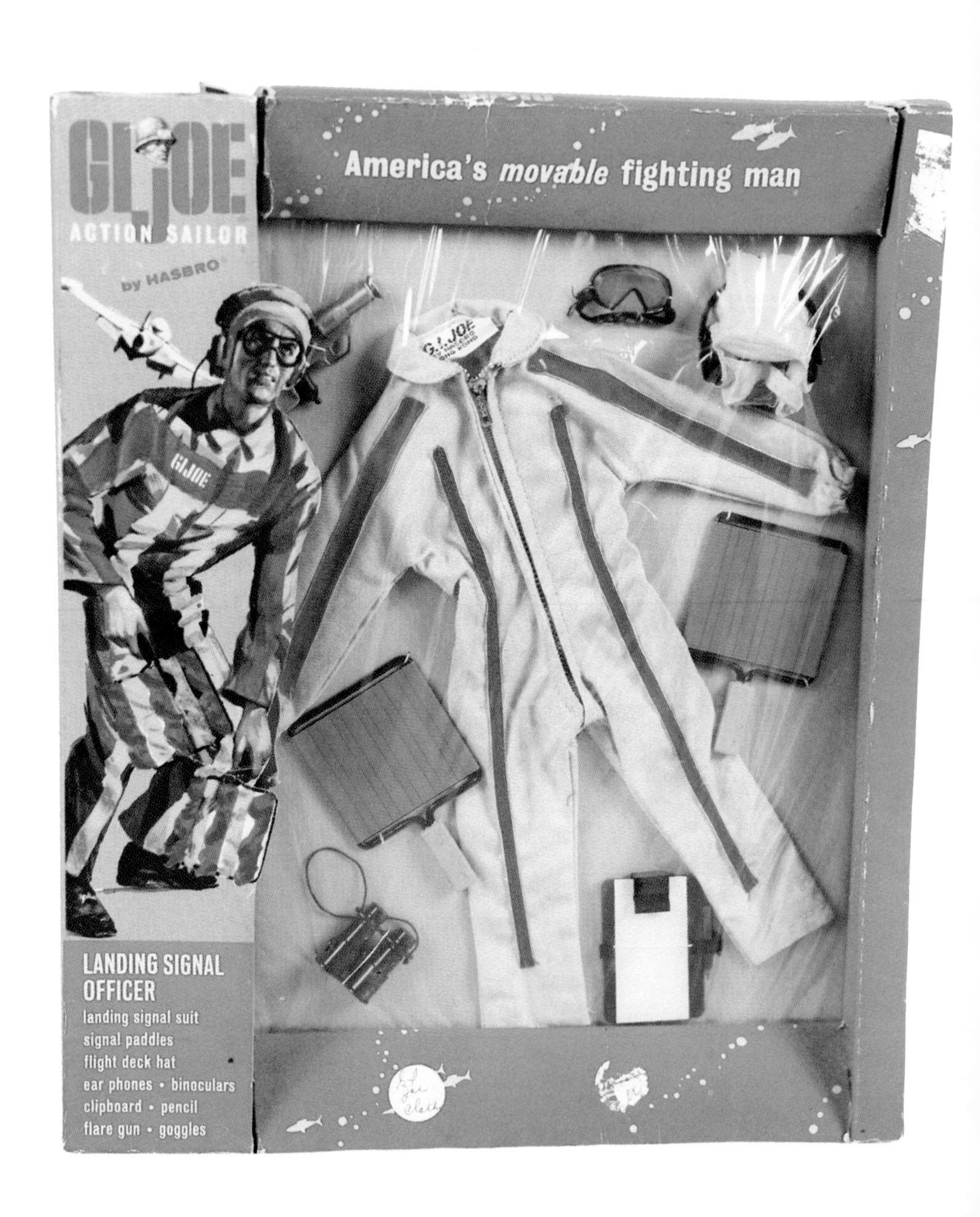

31. Landing Signal Officer Set.

32. Navy Attack Set.

33. Stock number 8103 *French Resistance Fighter* from "Action Soldiers of the World" series, painted black hair, blue eyes, dressed in black knit sweater, black pants, black boots, black plastic beret, Lebel revolver, shoulder holster, radio set with earphones, 7.65 MAS machine gun, knife, grenades, Croix de Guerre medal, counter-intelligence manual. 1966-1967. *Lila Ayala Collection.*

34. Stock number 8104 *British Commando* from "Action Soldiers of the World" series dressed in dark green flannel jacket and pants, brown boots, green helmet, gas mask, Sten Mark 2S sub-machine gun and case, canteen and cover, Victoria Cross medal, counter-intelligence manual. 1966-1967.

35. Stock number 8105 *Australian Jungle Fighter* from "Action Soldiers of the World" series, painted blonde hair, blue eyes, dressed in khaki bush jacket and shorts, knee-length olive green socks and brown boots, grenades, flamethrower, jungle knife, machete and sheath, entrenching tool, Victoria Cross medal, counter-intelligence manual. 1966-1967. *Lila Ayala Collection.*

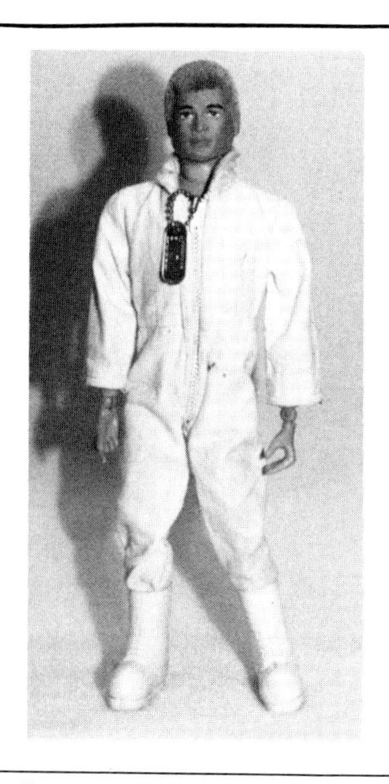

36. Stock number 1915 *Talking Astronaut,* blonde flocked hair, brown painted eyes, dressed in white coveralls, white boots, dog tag, insignia set, training manual. 1969-1970. *Lila Ayala Collection.*

37. Stock number 7400 *Talking Adventure Team Commander,* brown flocked hair and beard, painted blue eyes, dressed in green jacket and pants, black boots, pistol, shoulder holster, insignia set. 1970.

38. Stock number 7401, *Land Adventurer,* brown flocked hair and beard, painted blue eyes, dressed in camouflage fatigues, black boots, pistol, holster, dog tag, insignia, manual. 1970.

39. Shore Patrol Set.

40. Dress Parade Set.

41. Stock number 7404 *Black Adventurer* with flocked black hair, no beard, painted brown eyes, dressed in tan shirt and pants, black boots, pistol, shoulder holster, dog tag, insignia set. 1970.

42. Stock number 7500 *Man of Action* with flocked hair, no beard, painted blue eyes, dressed in green army fatigues, green cap, black boots, dog tag, insignia set, manual. 1970. *Lila Ayala Collection.*

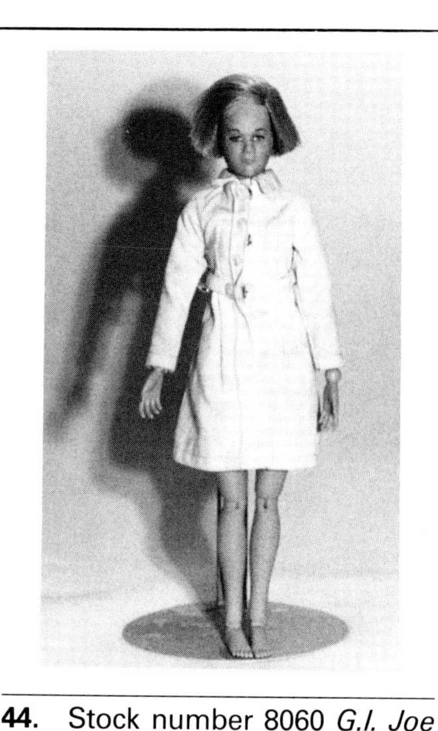

44. Stock number 8060 *G.I. Joe Action Nurse* with blonde rooted hair and eyelashes, painted green eyes, dressed in white nurse's uniform, white cap, white shoes, medical bag, stethoscope, plasma bottle, two crutches, package of bandages, splints. 1967.

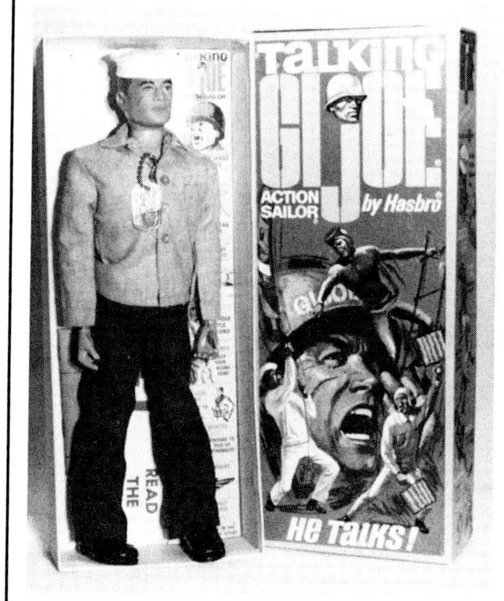

43. Stock number 7690 *Talking Sailor* with painted red hair, blue eyes, dressed in blue work shirt and pants, white cap, black boots, dog tag, insignia set, manual. 1967.

45. Medic Equipment and Gear.

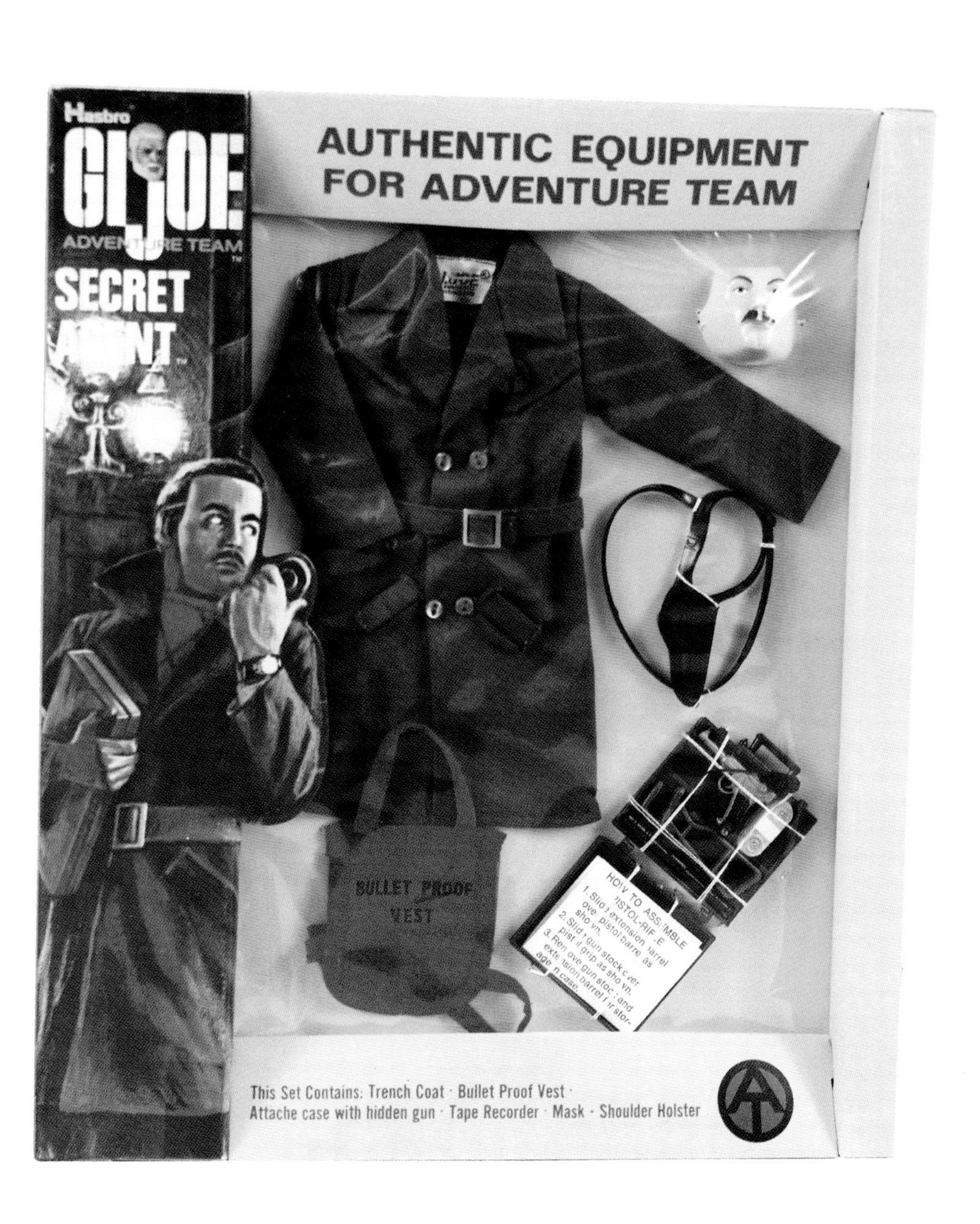

46. Secret Agent Equipment and Gear for the Adventure Team.

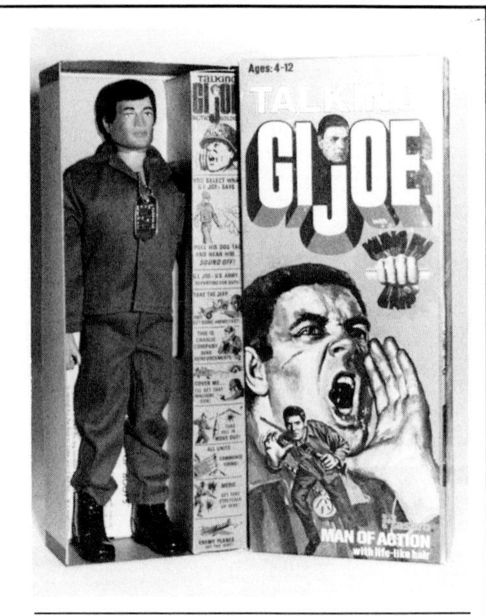

48. Stock number 7292 *Talking G.I. Joe* with Kung Fu Grip hands, flocked hair, no beard, blue painted eyes, dressed in green fatigues, black boots, dog tag, insignia set, rifle manual. 1974.

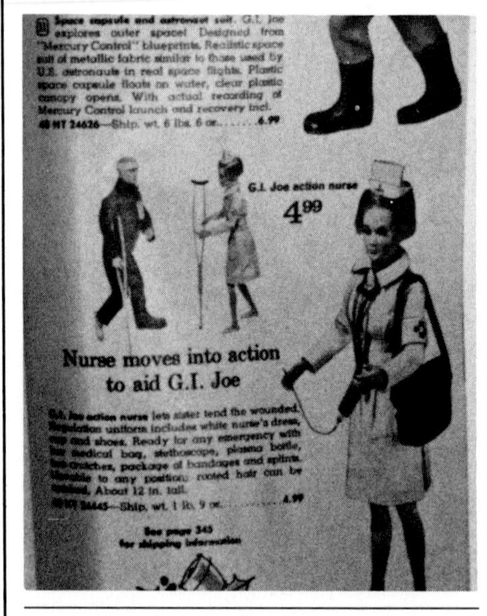

47. Stock number 8060 *G.I. Joe Action Nurse,* illustration from Montgomery Ward Christmas catalog. 1967.

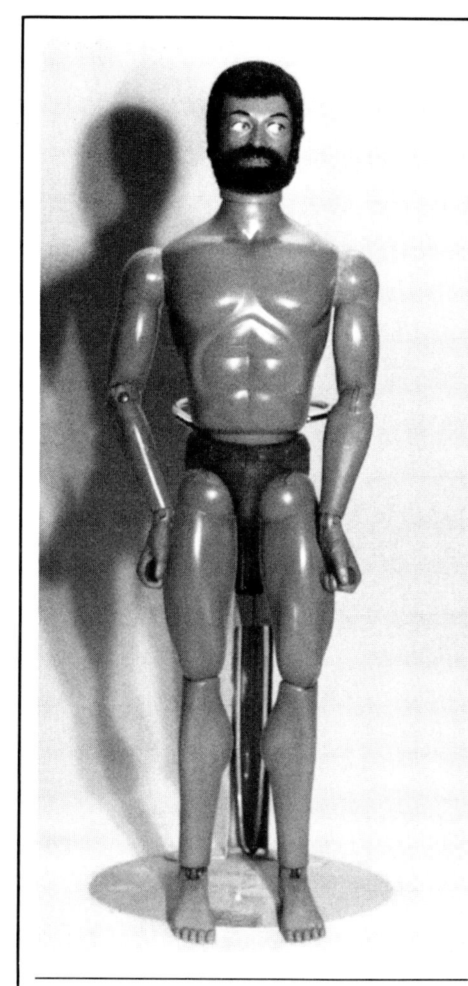

49. Stock number unknown, *Eagle Eye G.I. Joe* with flocked hair and beard, lever on back of head moves eyes from side to side, molded blue shorts. 1975-1977.

50. Stock Number 7270 *Adventure Team* with flocked hair and beard, painted blue eyes, molded blue shorts, dressed in camouflaged shorts only. 1975-1977.

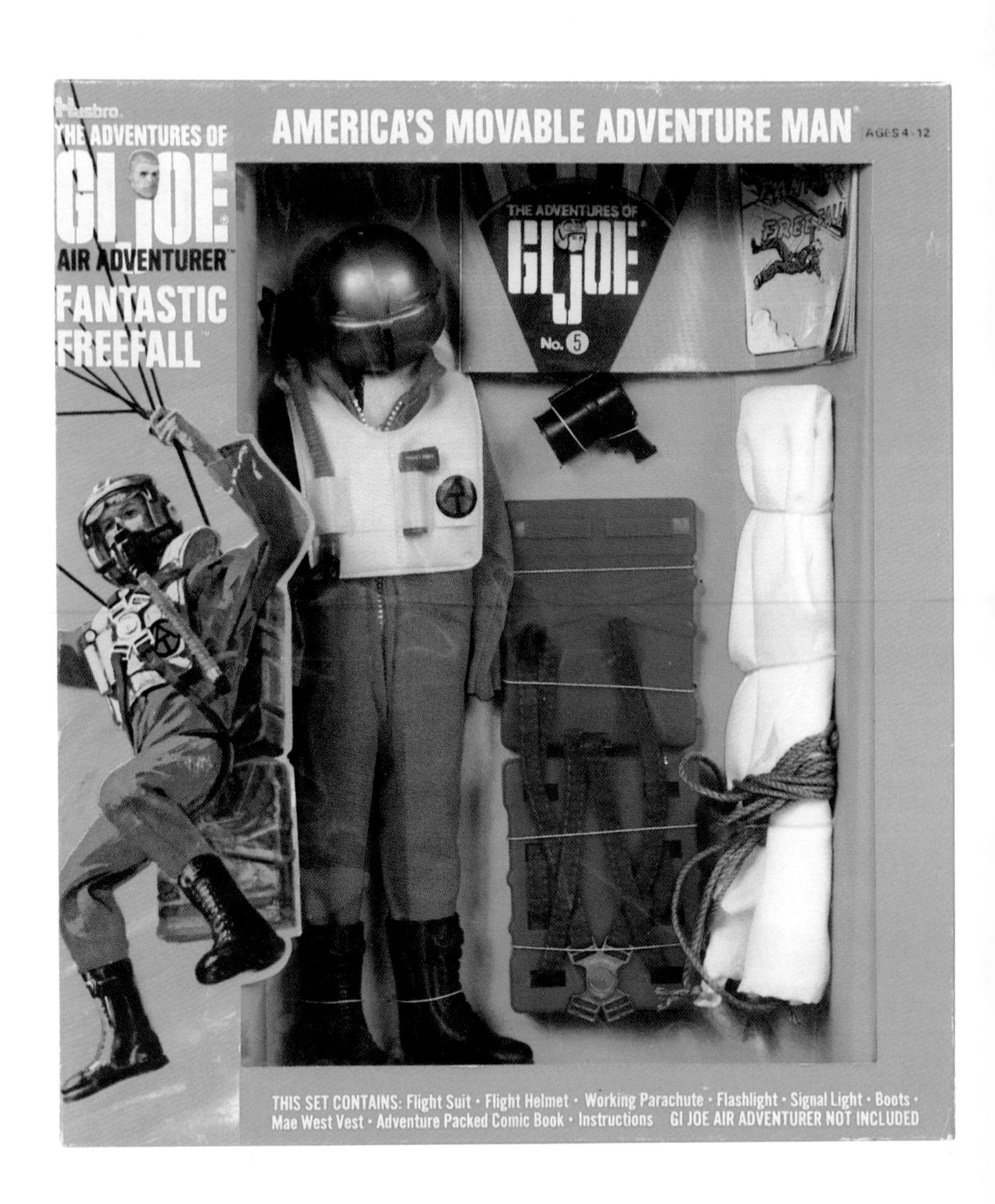

51. Fantastic Freefall Set.

52. Karaté Uniform Set.

53. Stock number 8026 *Bulletman;* both arms are silver, also both hands, painted brown hair and eyes, molded blue shorts, dressed in red sleeveless bodysuit, red boots, silver bullet helmet. 1977.

54. Stock number 8025 *Atomicman* (Mike Power) with painted hair and eyes, Kung Fu Grip hands, left leg and right arm are clear robot-like limbs, see-through eye, dressed in camouflaged shirt, brown shorts. Shirt is tagged "Atomicman." 1975.

56. Stock number 7501 Combat Field Jacket Set, M 1 rifle, bayonet, Army field jacket, six hand grenades, cartridge belt, Army manual. *Lila Ayala Collection.*

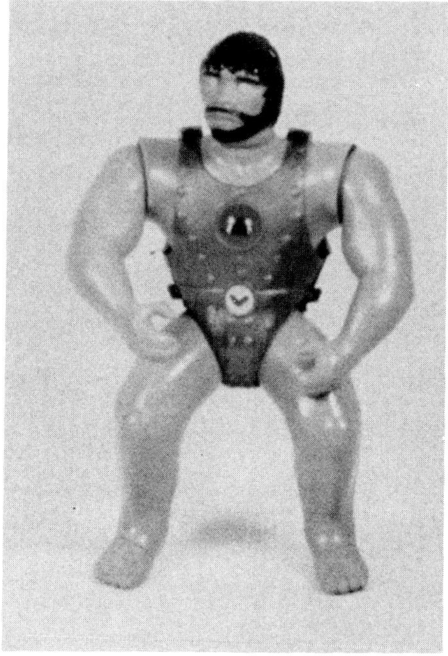

55. Stock number unknown, *Intruder,* squatty gorilla-type body, white eyes, button on back to make his arms grip, painted hair and beard, gold armor-styled bodysuit. 1976.

57. Gear and Equipment for Eight Ropes of Danger.

58. *Bulletman.*

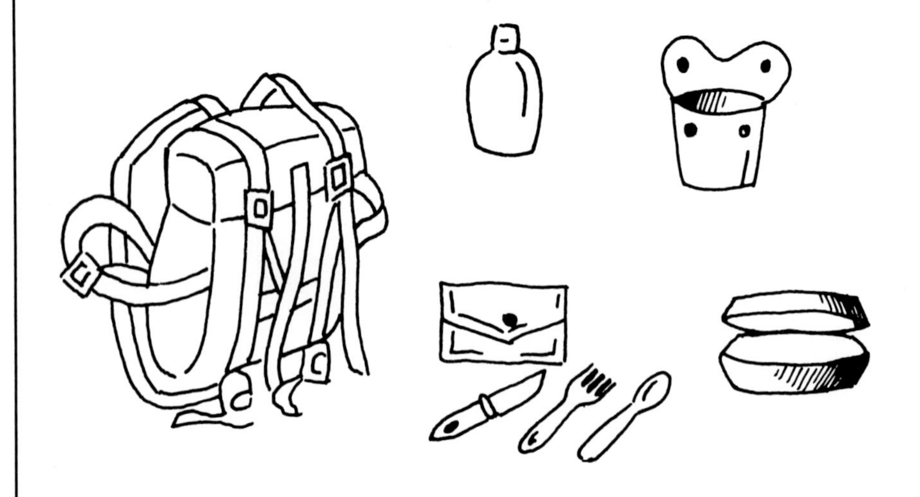

59. Stock number 7502 Combat Field Pack Set, field pack, mess kit, knife, fork, spoon, canteen, first-aid pouch, entrenching tool and cover, manual.

60. Stock number 7503 Combat Fatigue Shirt.

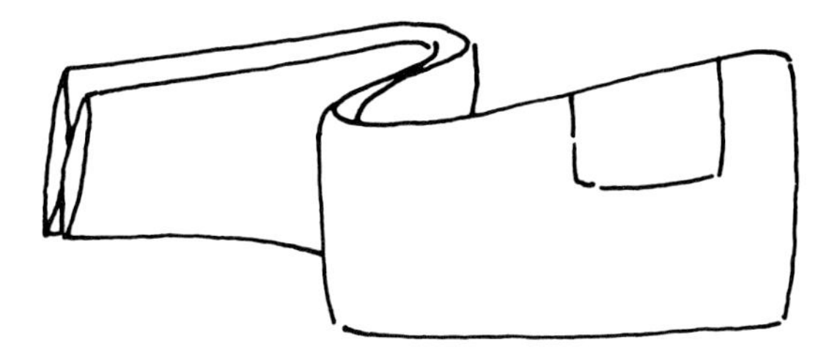

61. Stock number 7504 Combat Fatigue Pants.

62. Stock number 7505 Combat Field Jacket.

63. *Atomicman.*

64. Jaws of Death Gear and Equipment.

65. Stock number 7506 Combat Field Pack Set, field pack with entrenching tool and cover.

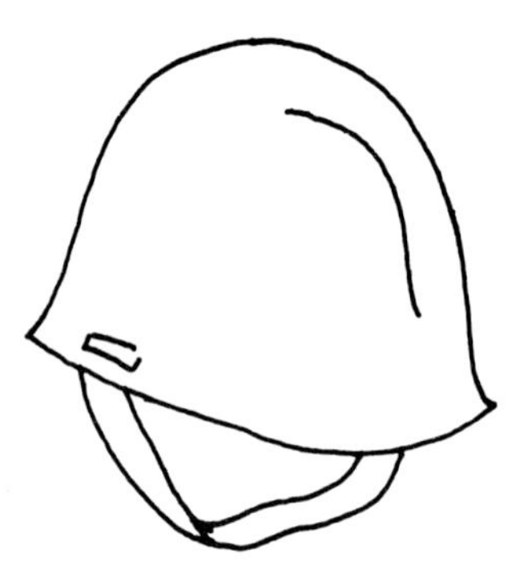

66. Stock number 7507 Combat Helmet Set, helmet with camouflage netting, foliage and leaves.

67. Stock number 7508 Combat Sandbag Set.

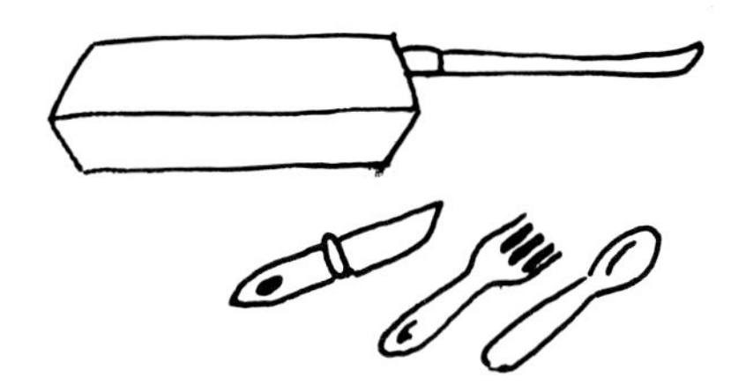

68. Stock number 7509 Combat Mess Kit Set, spoon, fork, knife, canteen with cover, mess kit.

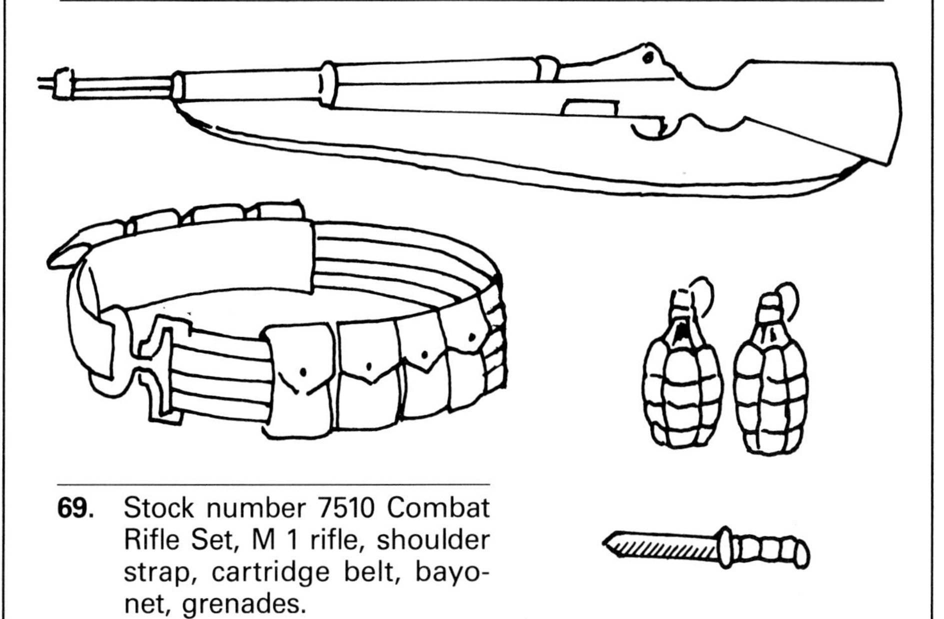

69. Stock number 7510 Combat Rifle Set, M 1 rifle, shoulder strap, cartridge belt, bayonet, grenades.

70. Stock number 7511 Combat Camouflage Netting Set, netting, foliage, posts.

71. *Atomicman* Secret Mountain Outpost Set.

72. White Tiger Hunt Set.

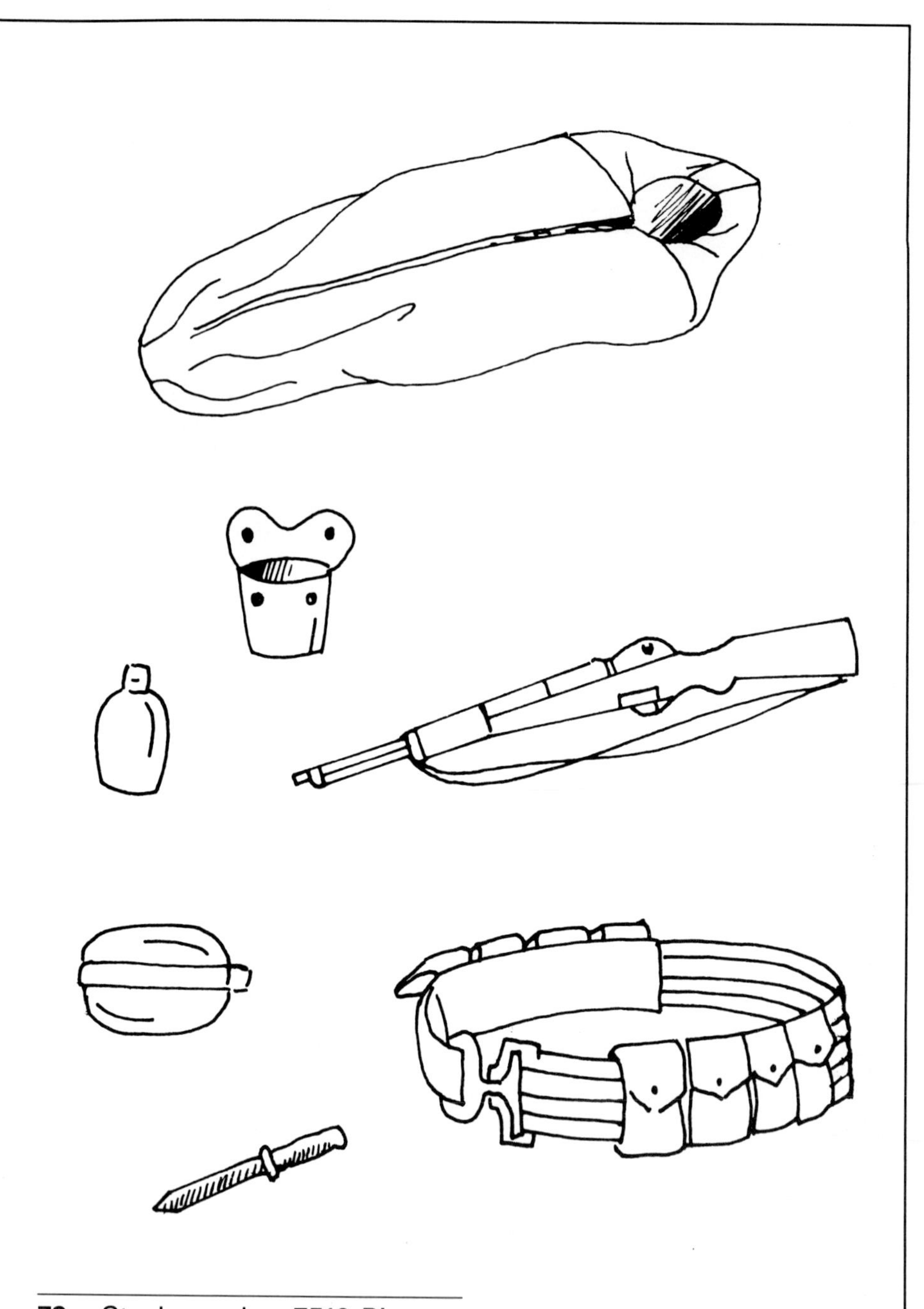

73. Stock number 7512 Bivouac Sleeping Bag Set, sleeping bag, M 1 rifle, bayonet, cartridge belt, mess kit, canteen and cover.

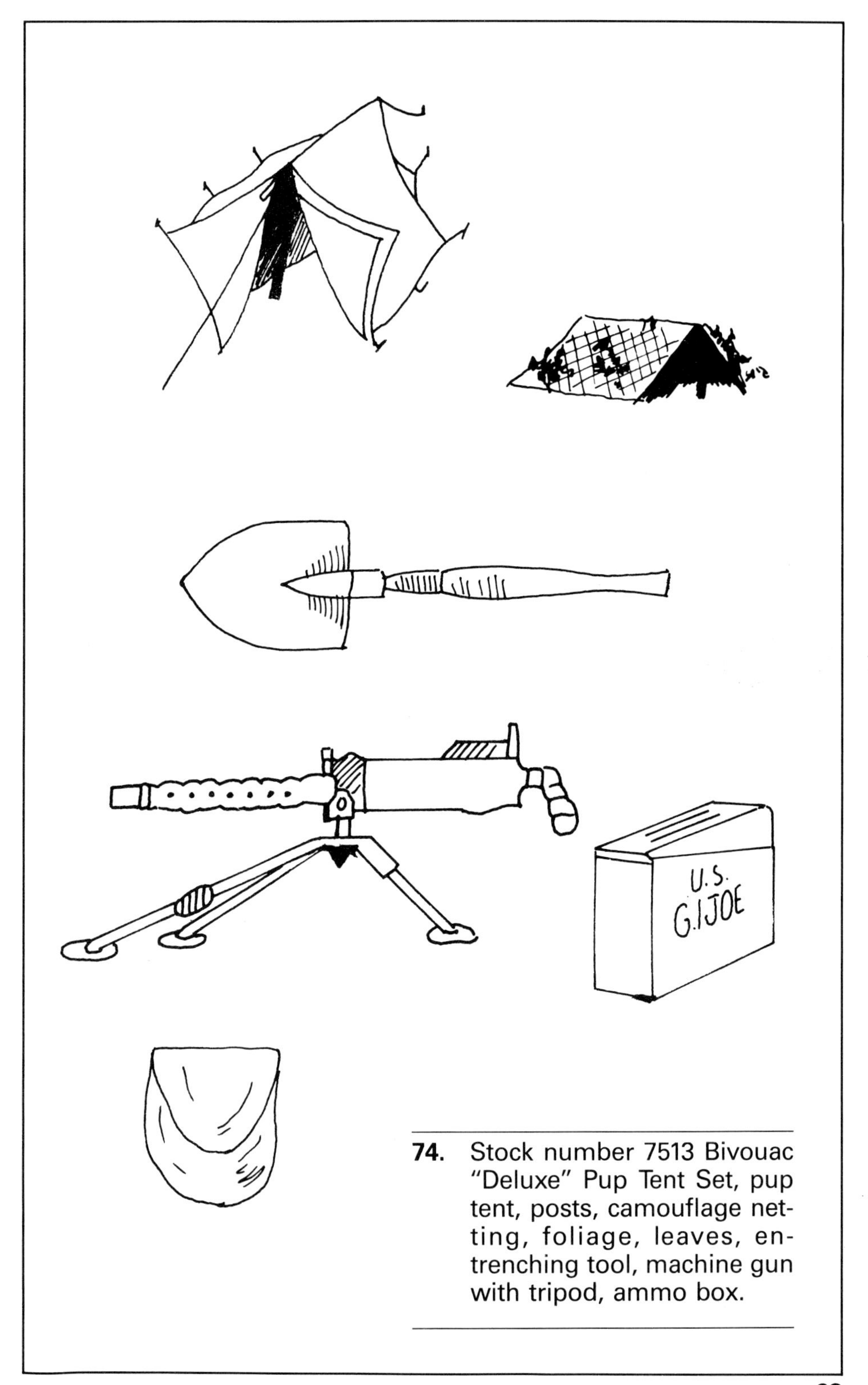

74. Stock number 7513 Bivouac "Deluxe" Pup Tent Set, pup tent, posts, camouflage netting, foliage, leaves, entrenching tool, machine gun with tripod, ammo box.

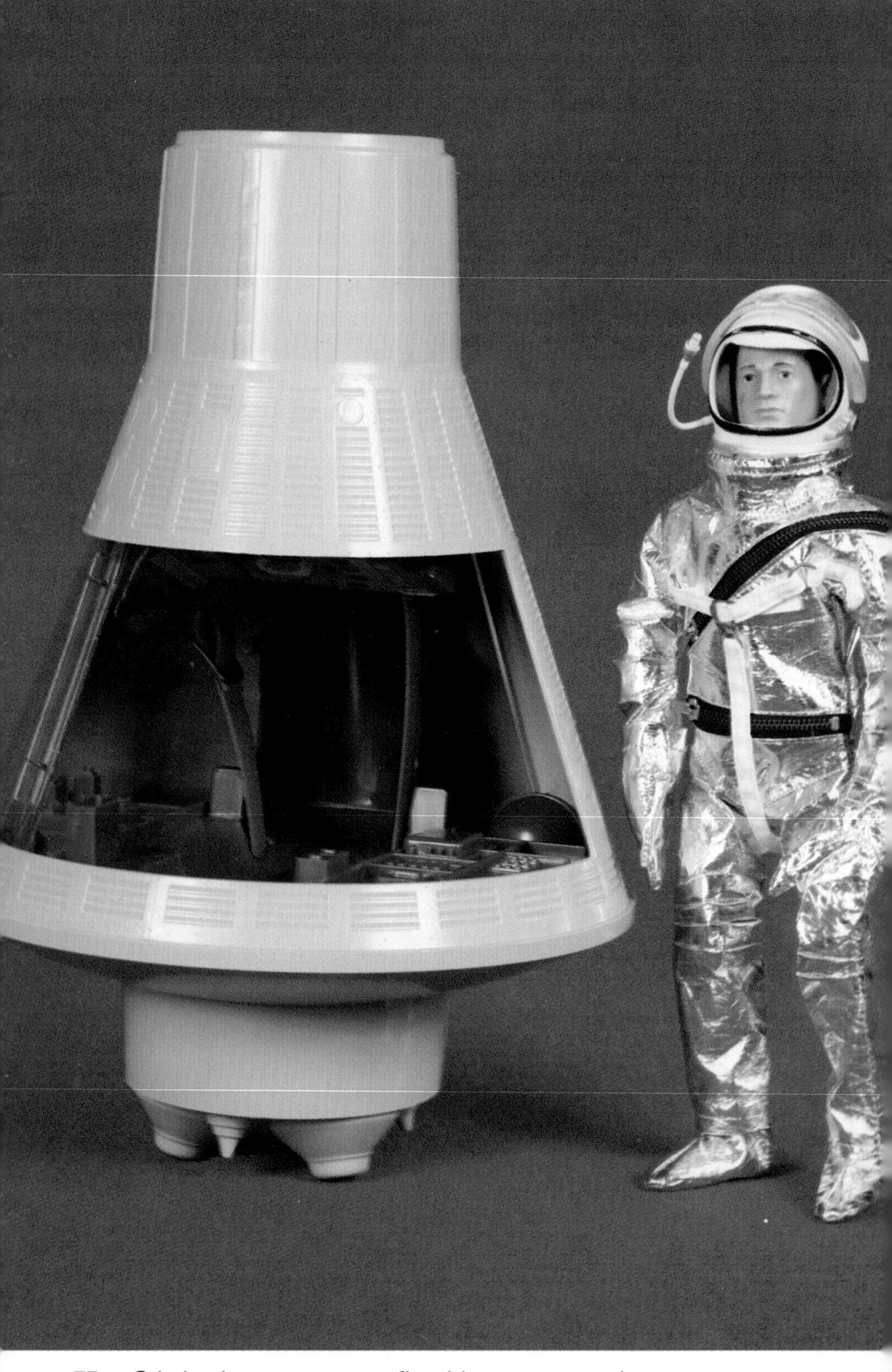

75. *G.I. Joe* in astronaut outfit with space capsule.

76. Stock number 7514 Bivouac Machine Gun, machine gun, tripod, ammo box.

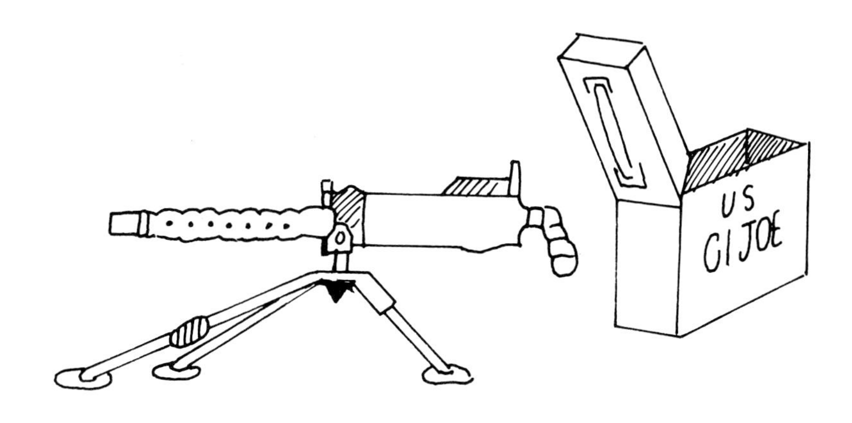

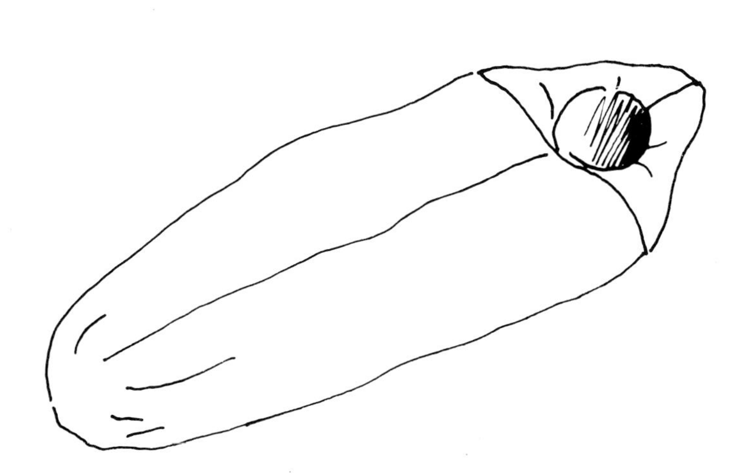

77. Stock number 7515 Bivouac Sleeping Bag.

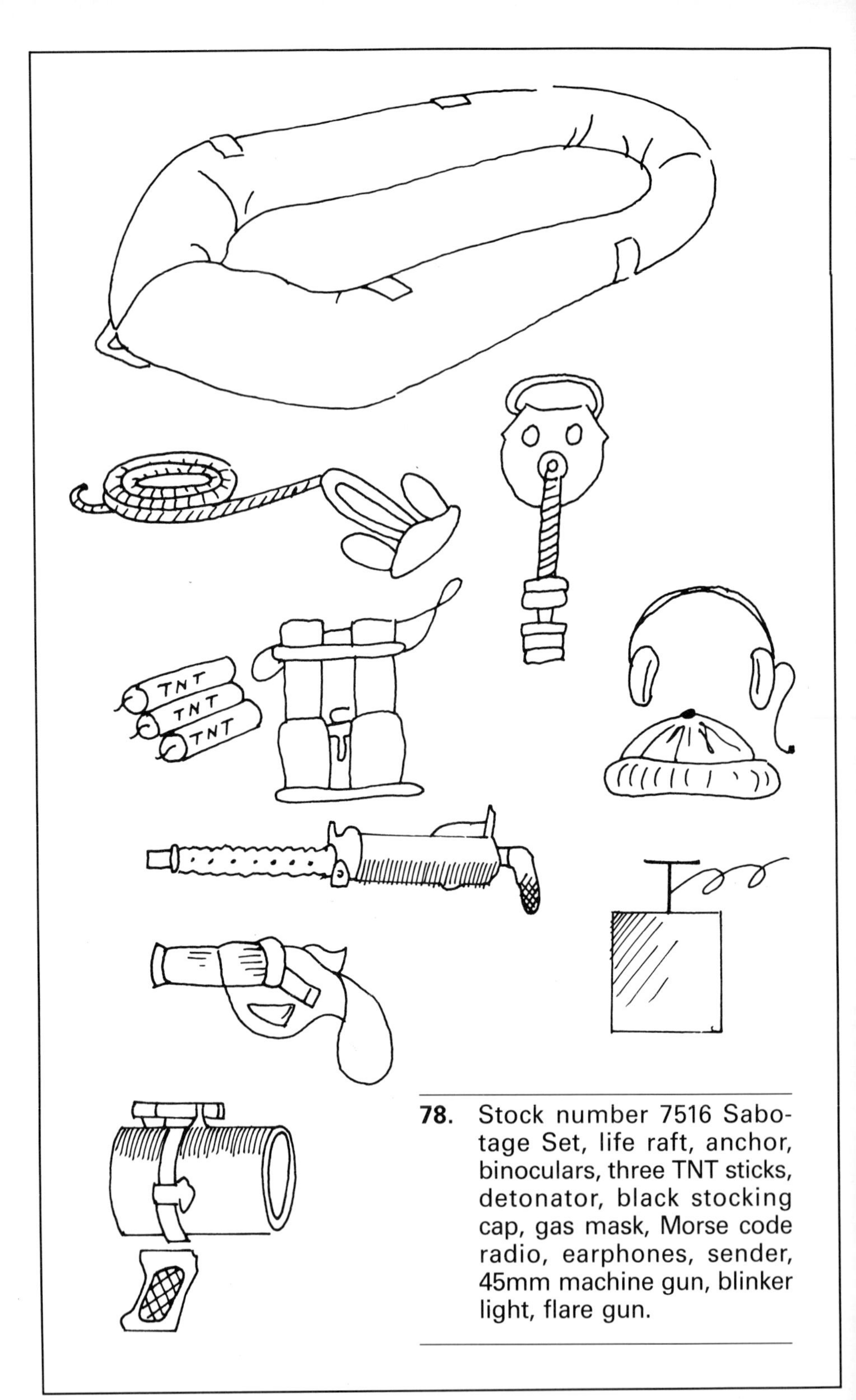

78. Stock number 7516 Sabotage Set, life raft, anchor, binoculars, three TNT sticks, detonator, black stocking cap, gas mask, Morse code radio, earphones, sender, 45mm machine gun, blinker light, flare gun.

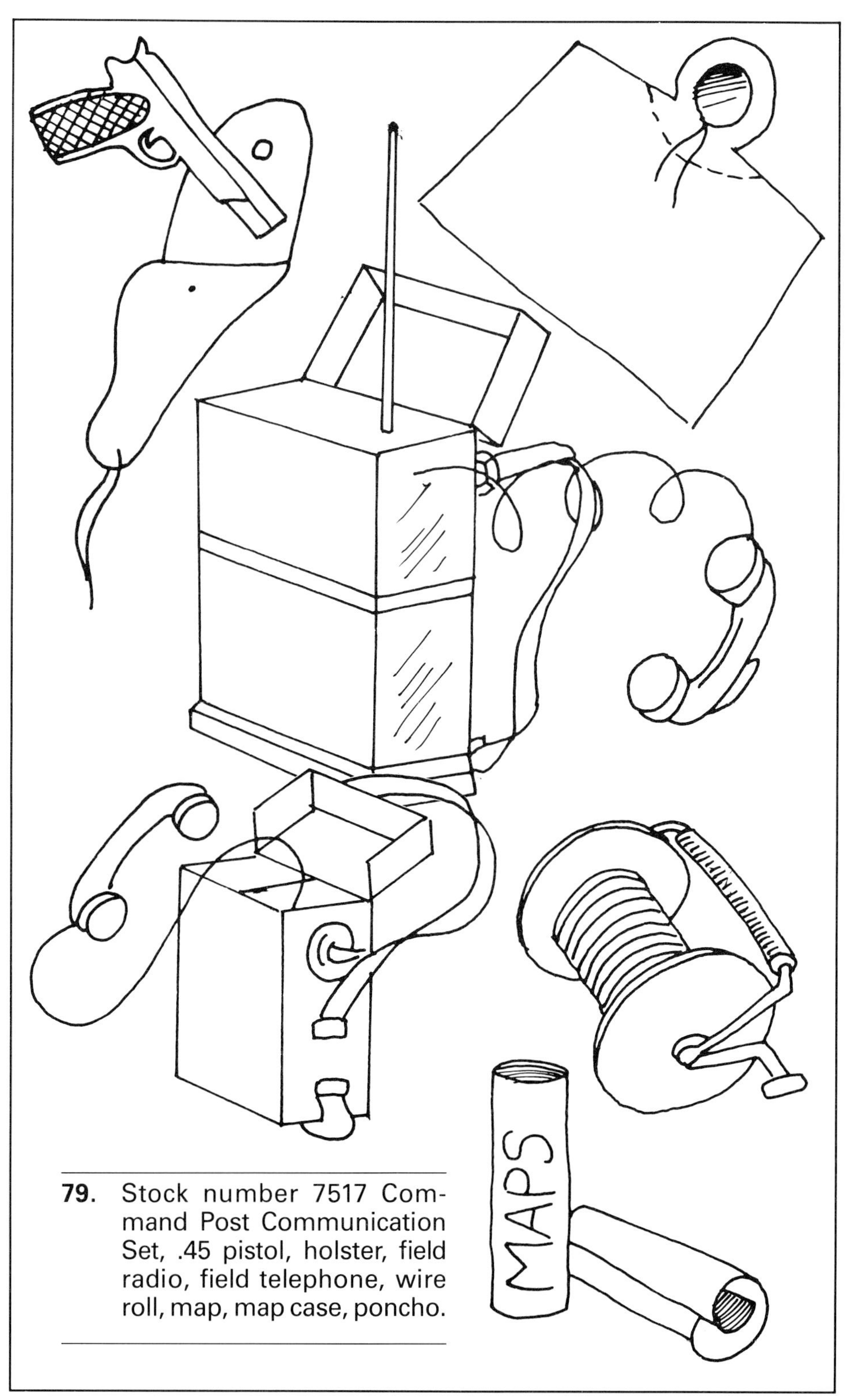

79. Stock number 7517 Command Post Communication Set, .45 pistol, holster, field radio, field telephone, wire roll, map, map case, poncho.

80. Stock number 7518 Command Post Small Arms Set, .45 pistol, holster, belt, first-aid pouch, grenades.

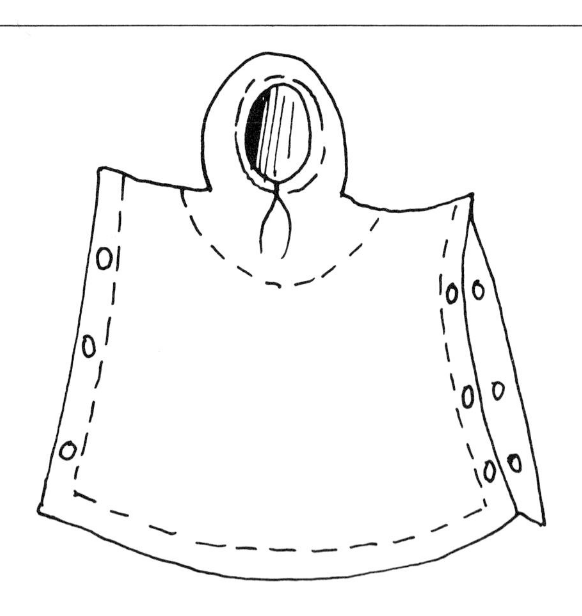

81. Stock number 7519 Command Post Poncho.

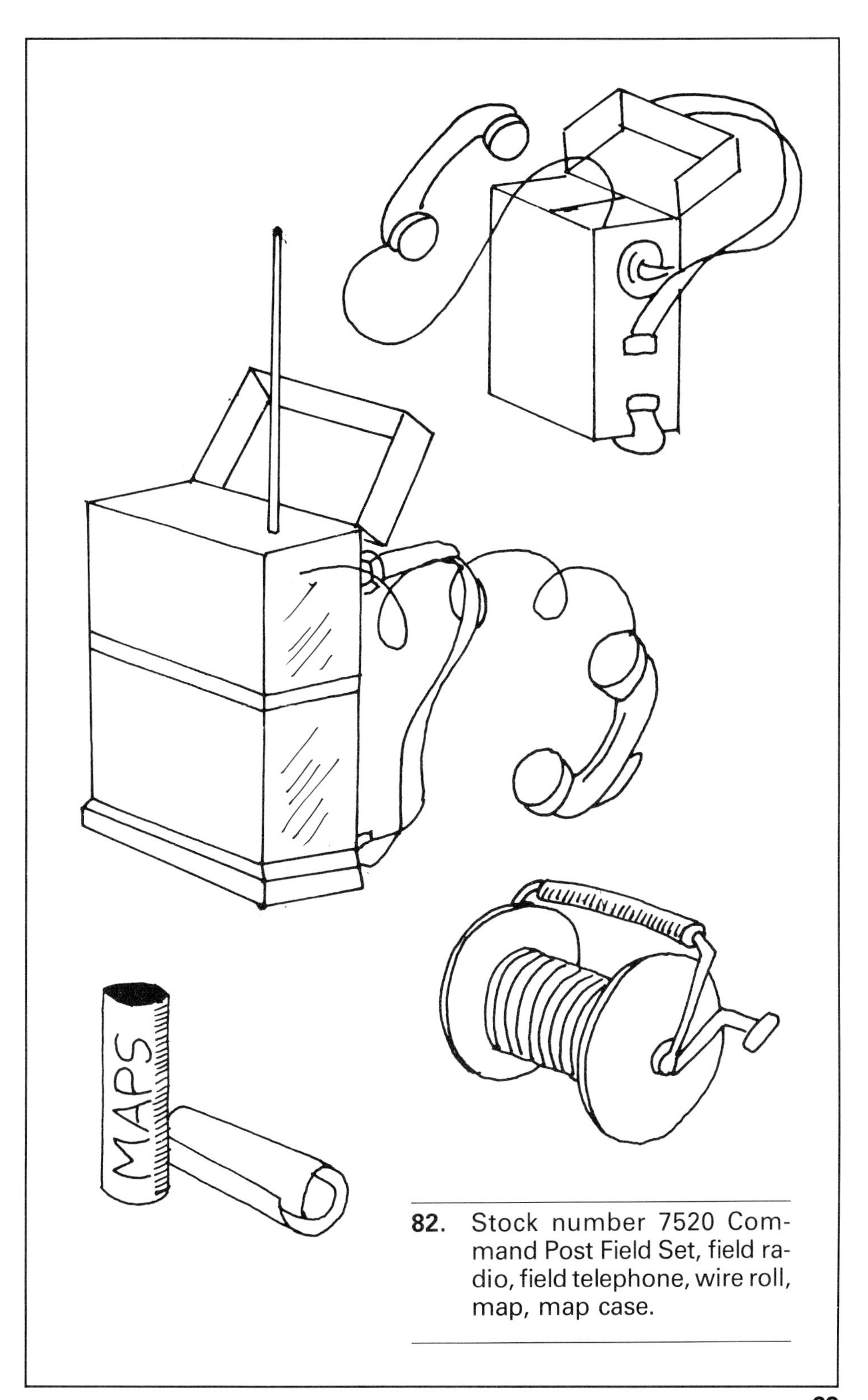

82. Stock number 7520 Command Post Field Set, field radio, field telephone, wire roll, map, map case.

83. Stock number 7521 Military Police Set, Ike jacket, pants, red scarf, billy club, .45 pistol, holster, belt, armband, duffle bag. *Lila Ayala Collection.*

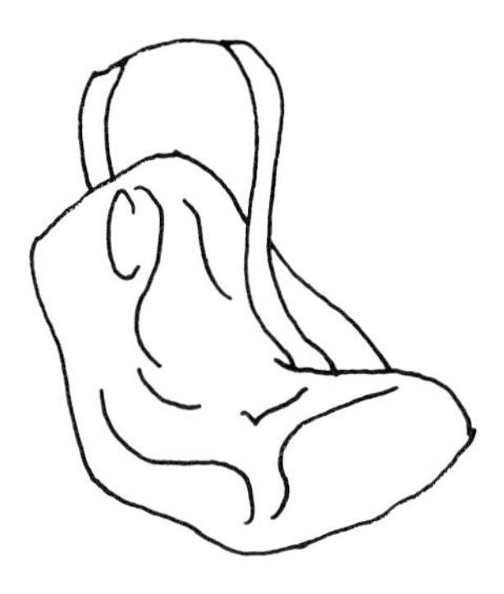

84. Stock number 7523 M.P. Duffle Bag.

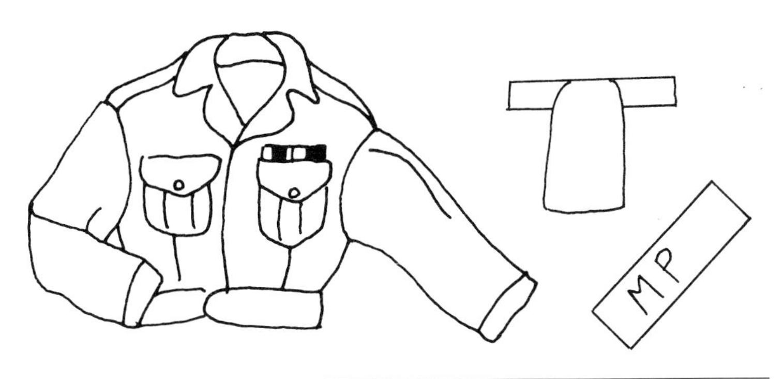

85. Stock number 7524 M.P. Ike Jacket Set, Ike jacket, red scarf, M.P. armband.

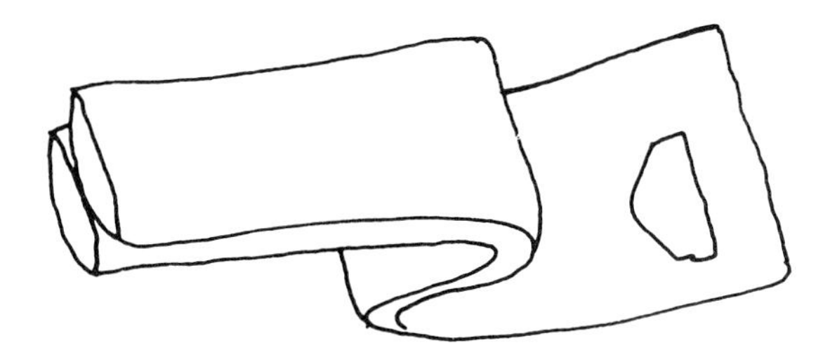

86. Stock number 7525 M.P. Ike Pants.

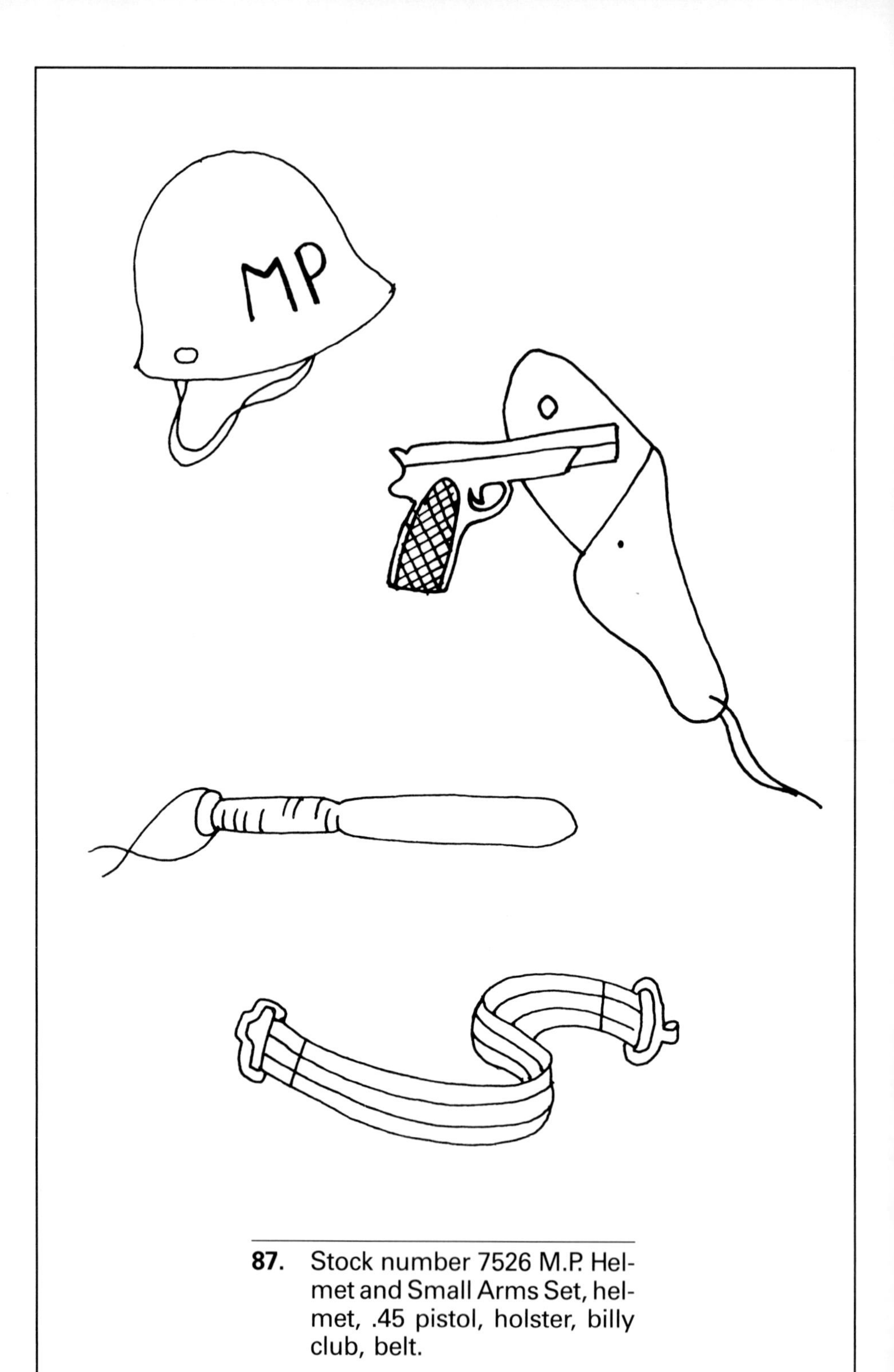

87. Stock number 7526 M.P. Helmet and Small Arms Set, helmet, .45 pistol, holster, billy club, belt.

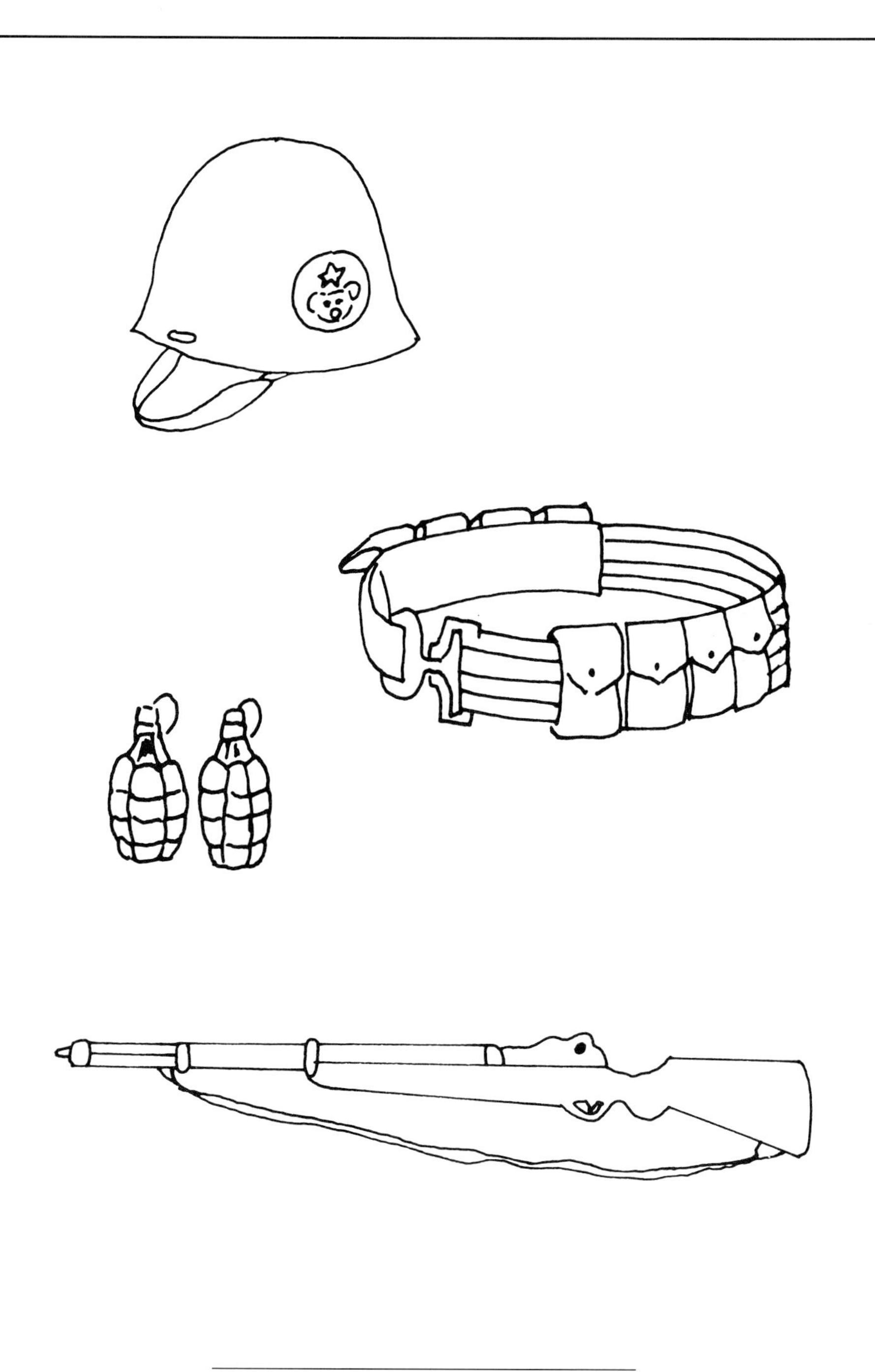

88. Stock number 7527 Ski Patrol camouflage winter white helmet, belt, grenades.

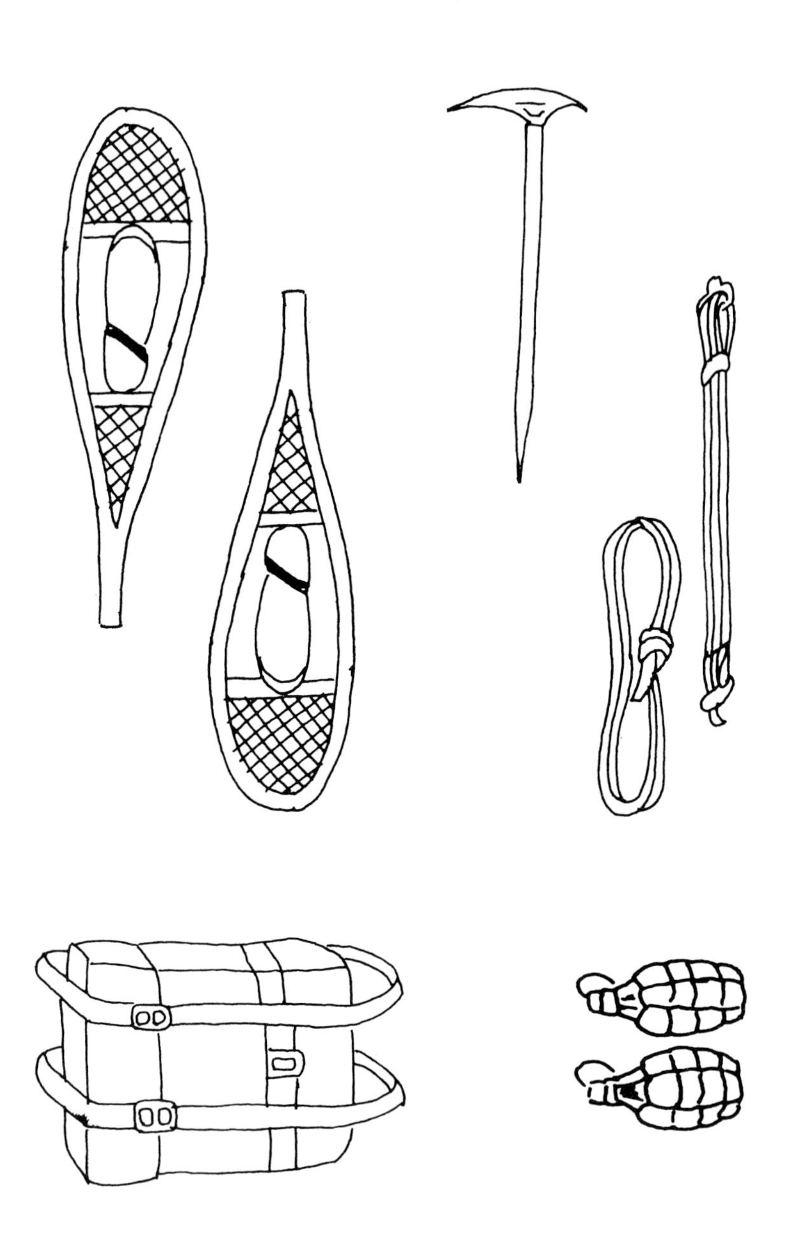

89. Stock number 7530 Mountain Troops winter white camouflage pack, web belt, snow shoes, ice ax, climbing rope, grenades.

90. Stock number 7531 Ski Patrol Set, white parka, pants, belt, mittens, skis, poles, goggles, white boots. *Lila Ayala Collection.*

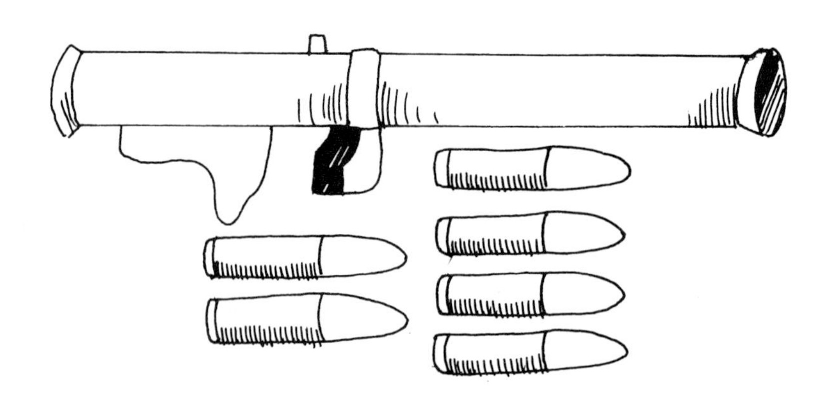

91. Stock number 7532 Special Forces Bazooka Set.

92. Stock number 7536 Green Beret and Equipment Set; camouflage shirt and pants, green plastic beret, black boots, bazooka, six rockets, grenades, machine gun, rifle, ammo box, field telephone, cartridge belt, camouflage netting. *Lila Ayala Collection.*

93. Stock number 7537 Westpoint Cadet Set gray jacket, white trousers, chest sash and buckle, white M 1 rifle, sword and scabbard, hat with plume, black shoes, waist sash. *Lila Ayala Collection.*

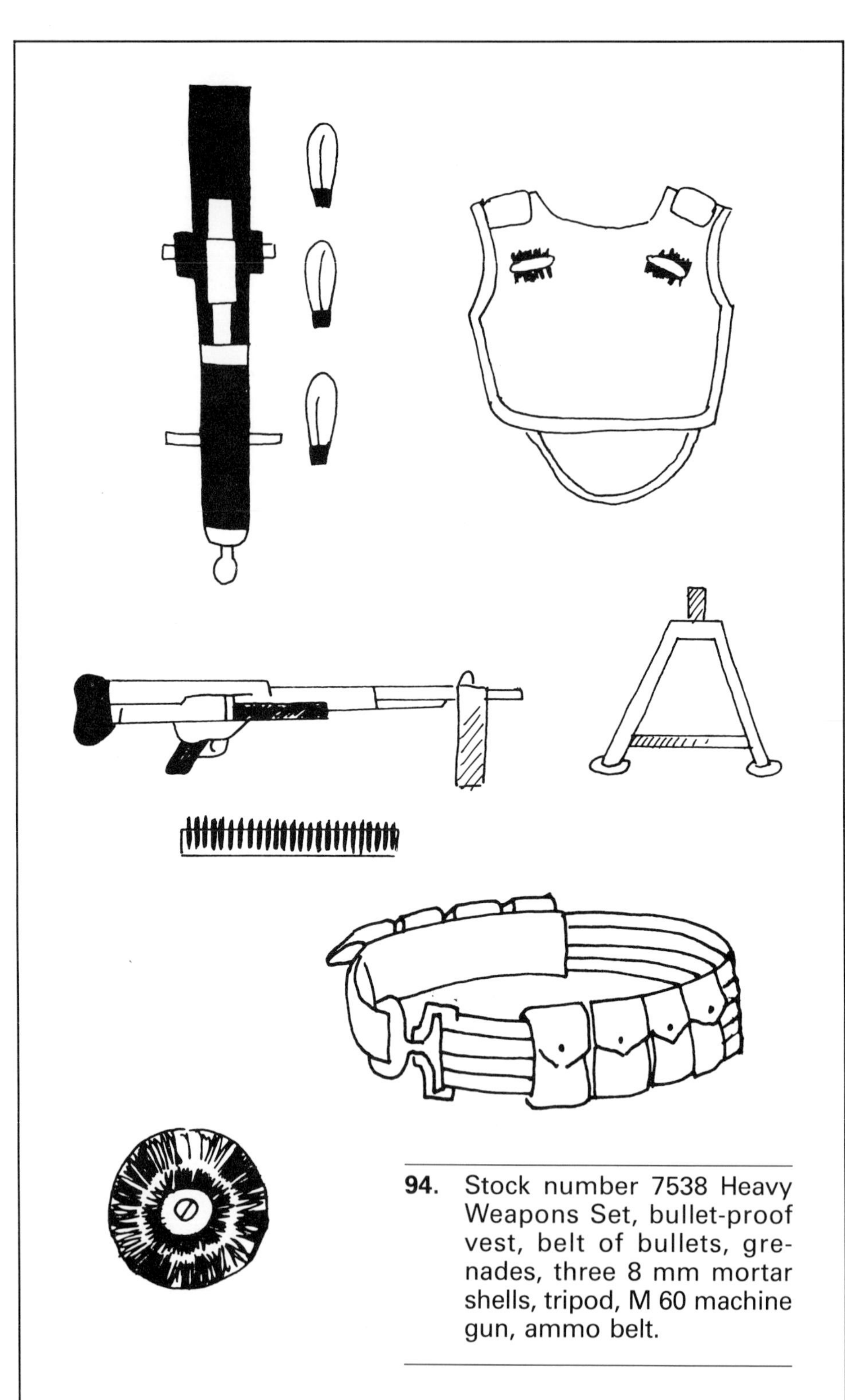

94. Stock number 7538 Heavy Weapons Set, bullet-proof vest, belt of bullets, grenades, three 8 mm mortar shells, tripod, M 60 machine gun, ammo belt.

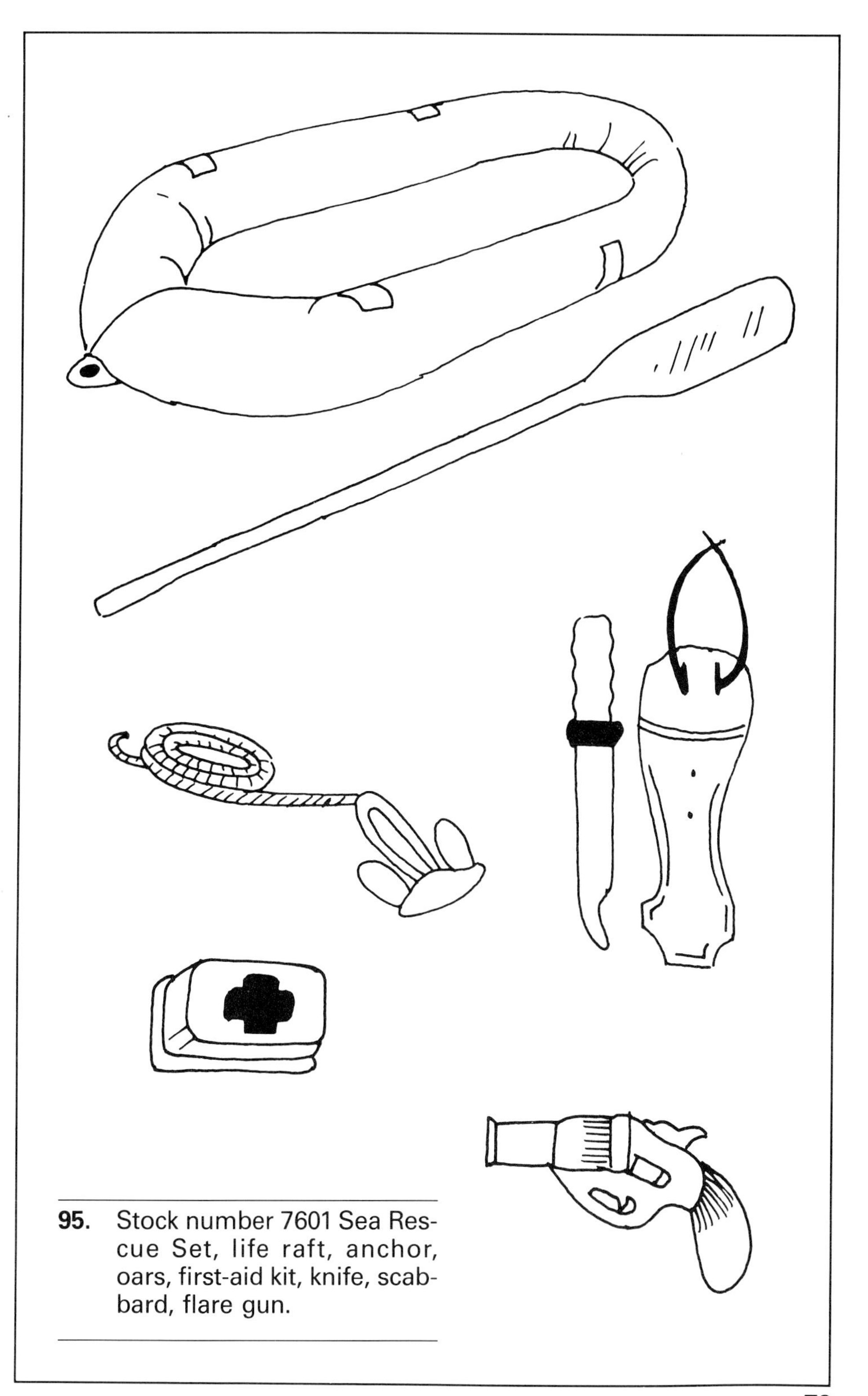

95. Stock number 7601 Sea Rescue Set, life raft, anchor, oars, first-aid kit, knife, scabbard, flare gun.

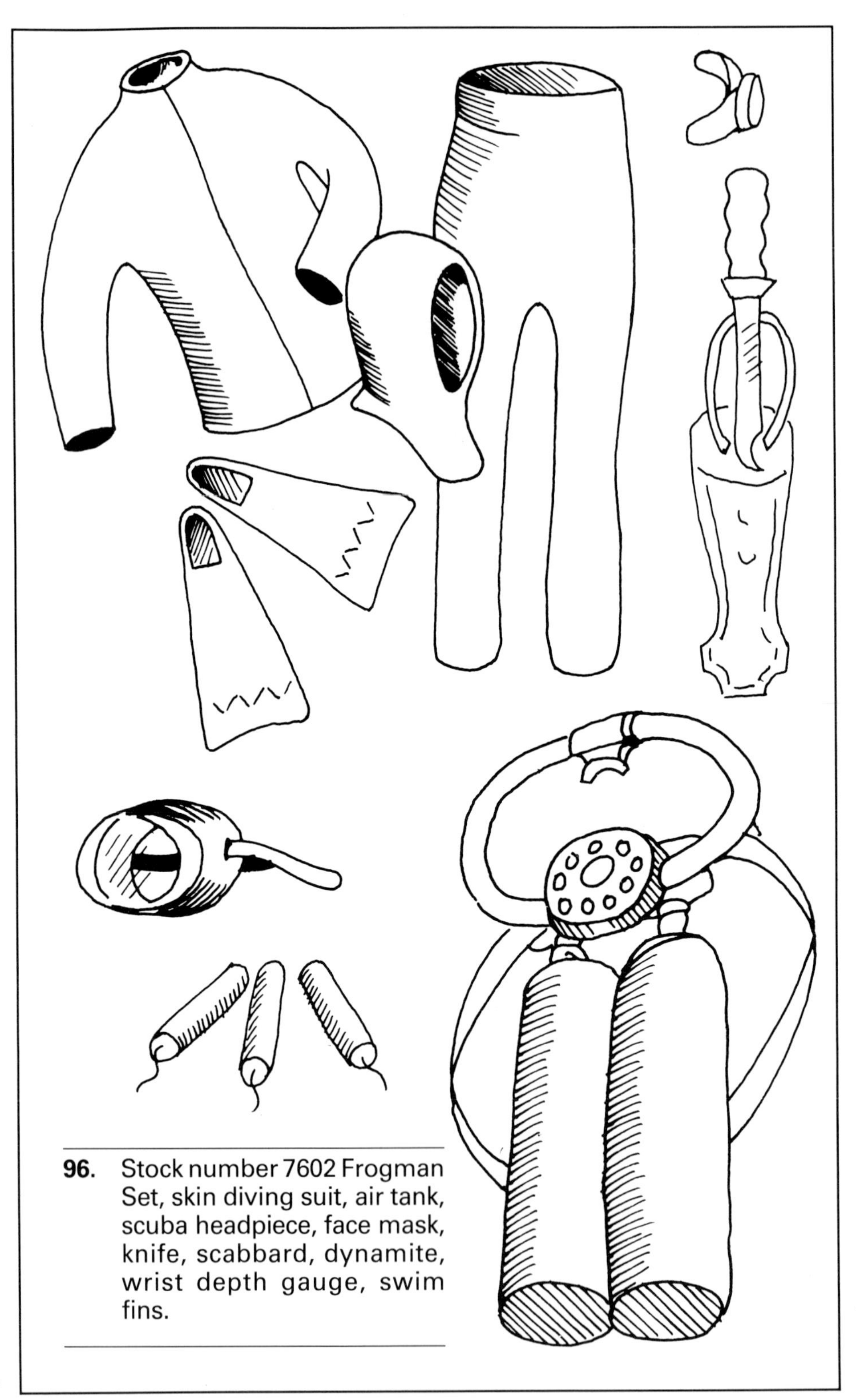

96. Stock number 7602 Frogman Set, skin diving suit, air tank, scuba headpiece, face mask, knife, scabbard, dynamite, wrist depth gauge, swim fins.

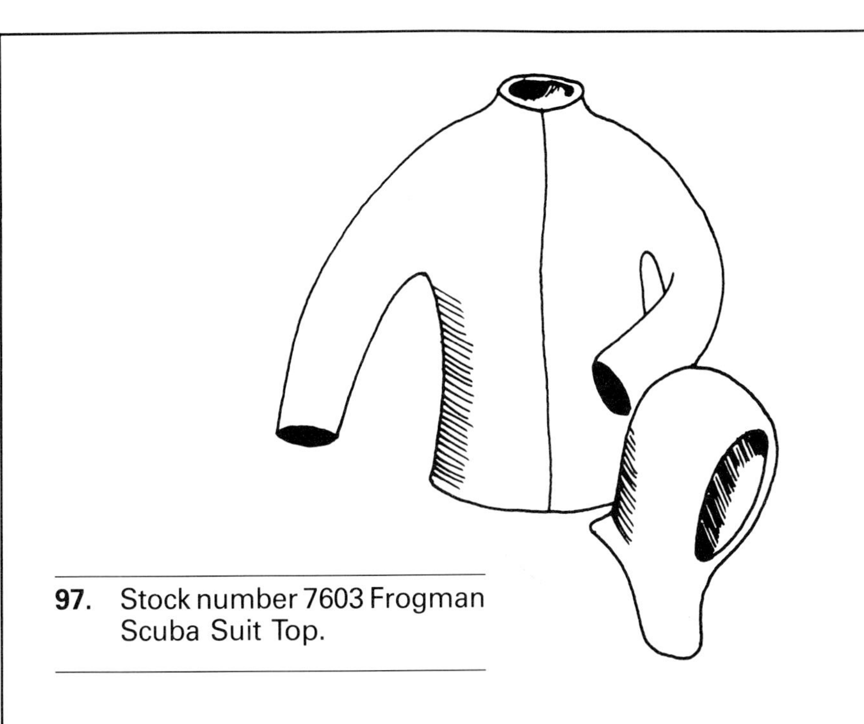

97. Stock number 7603 Frogman Scuba Suit Top.

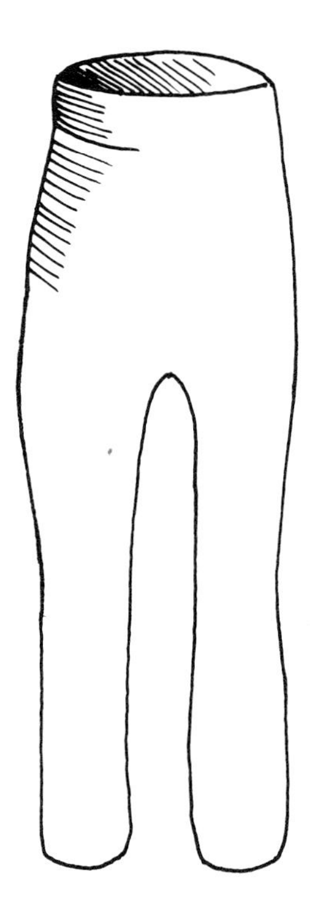

98. Stock number 7604 Frogman Scuba Suit Bottom.

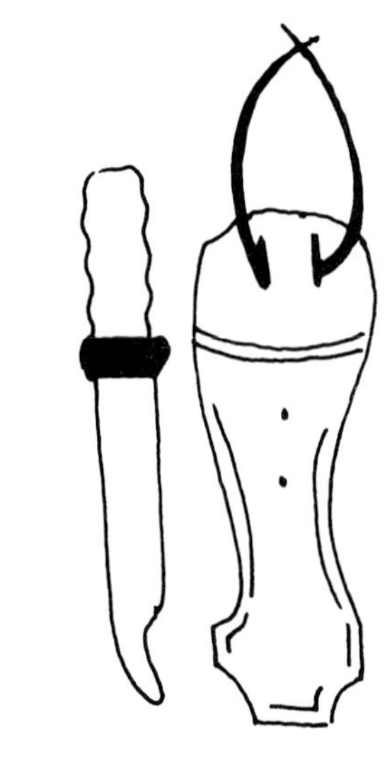

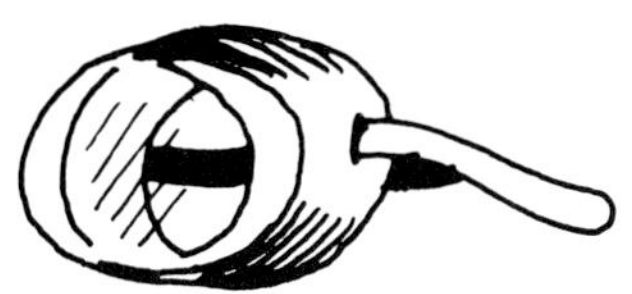

99. Stock number 7605 Frogman Set, face mask, fins, wrist gauge, knife, scabbard.

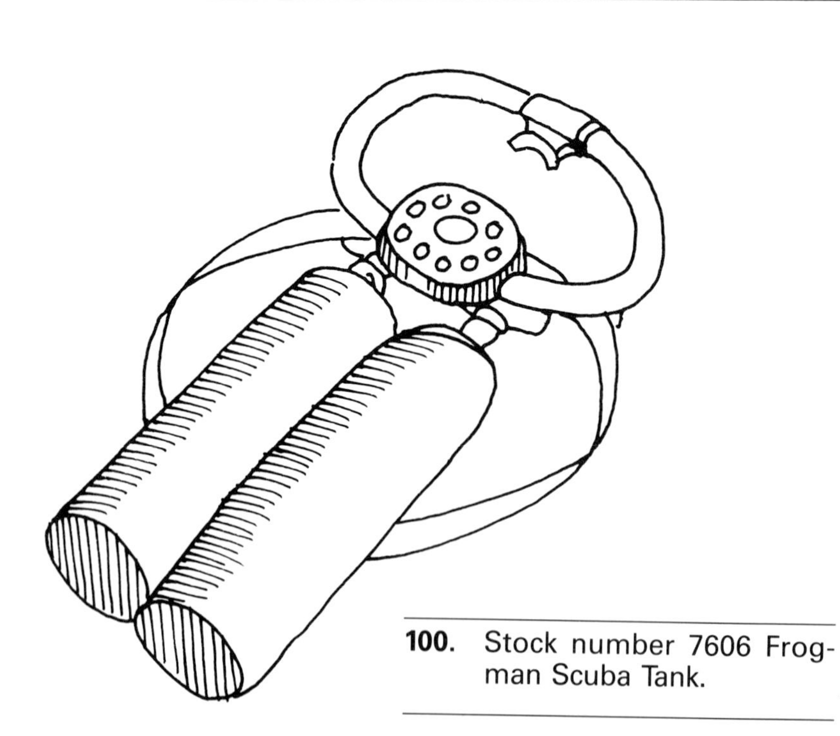

100. Stock number 7606 Frogman Scuba Tank.

101. Stock number 7607 Navy Attack Set, life jacket, blinker light, field glasses, semaphore flags.

102. Stock number 7608 Navy Attack Work Shirt.

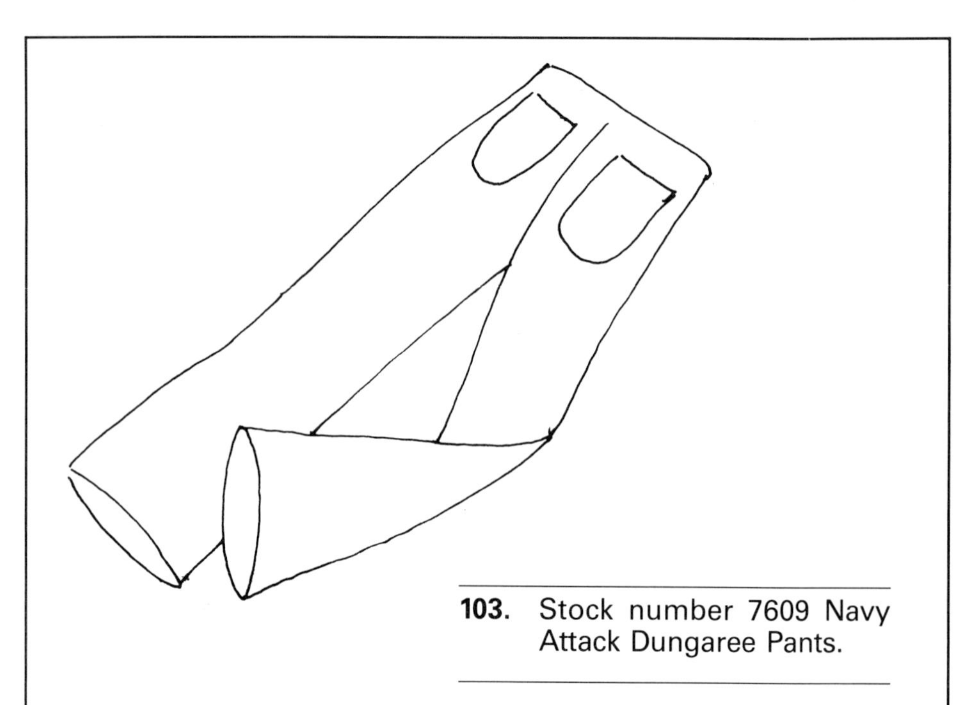

103. Stock number 7609 Navy Attack Dungaree Pants.

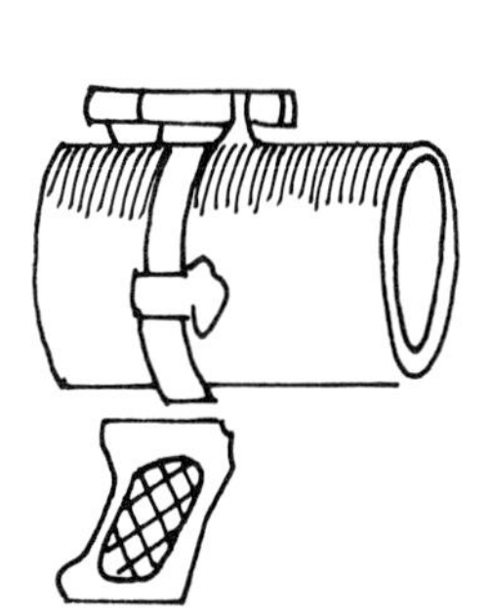

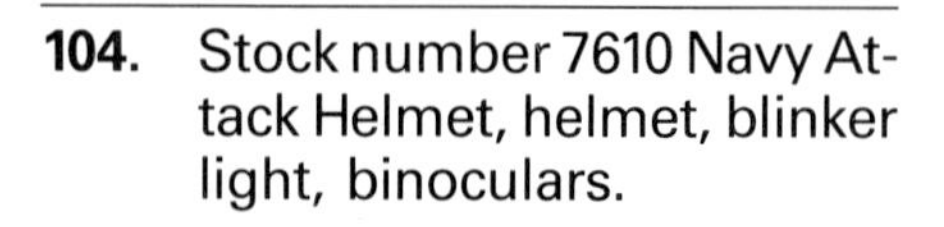

104. Stock number 7610 Navy Attack Helmet, helmet, blinker light, binoculars.

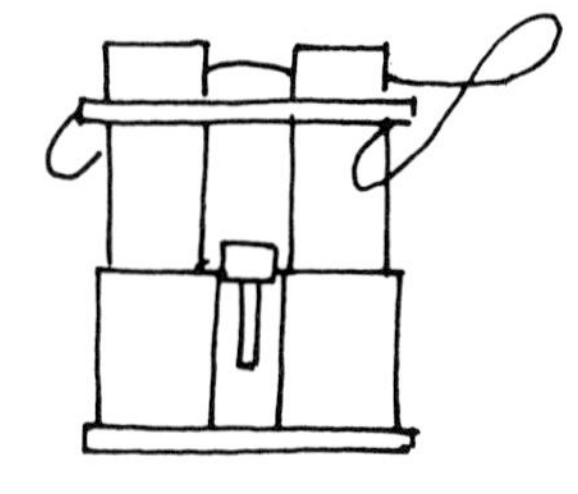

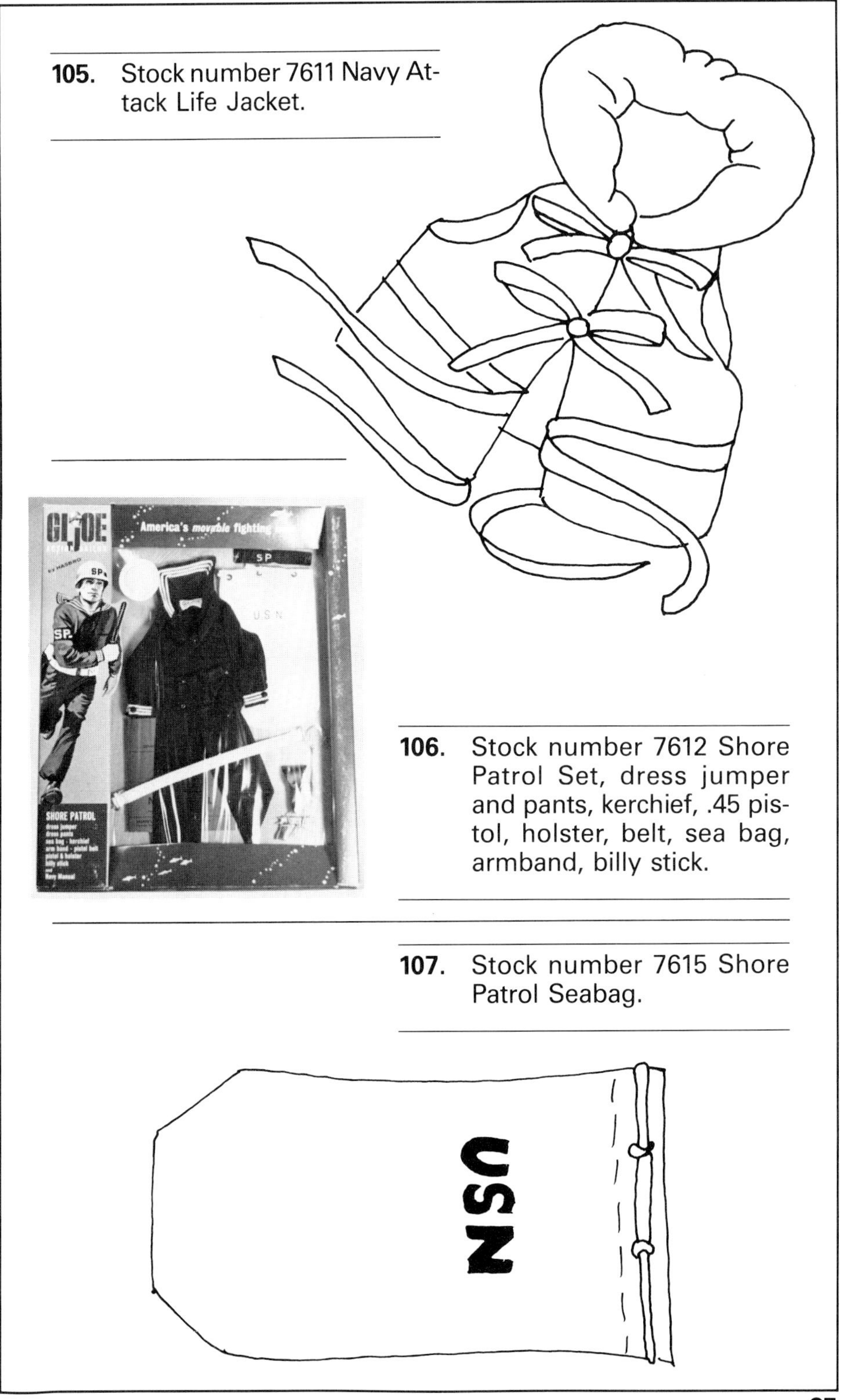

105. Stock number 7611 Navy Attack Life Jacket.

106. Stock number 7612 Shore Patrol Set, dress jumper and pants, kerchief, .45 pistol, holster, belt, sea bag, armband, billy stick.

107. Stock number 7615 Shore Patrol Seabag.

108. Stock number 7616 Shore Patrol Small Arms Set, helmet, billy club.

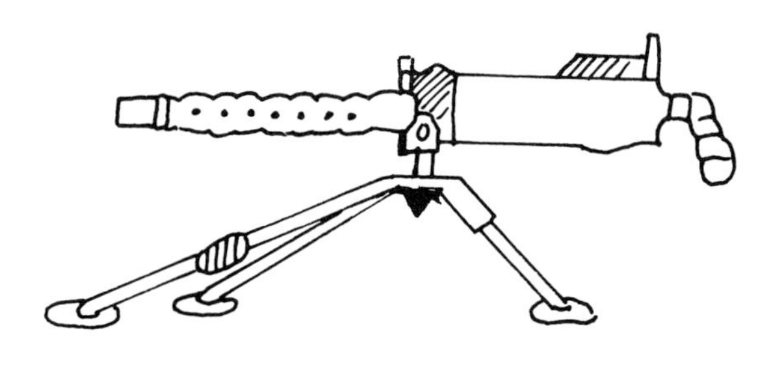

109. Stock number 7618 Machine Gun, 30 caliber machine gun, ammo box.

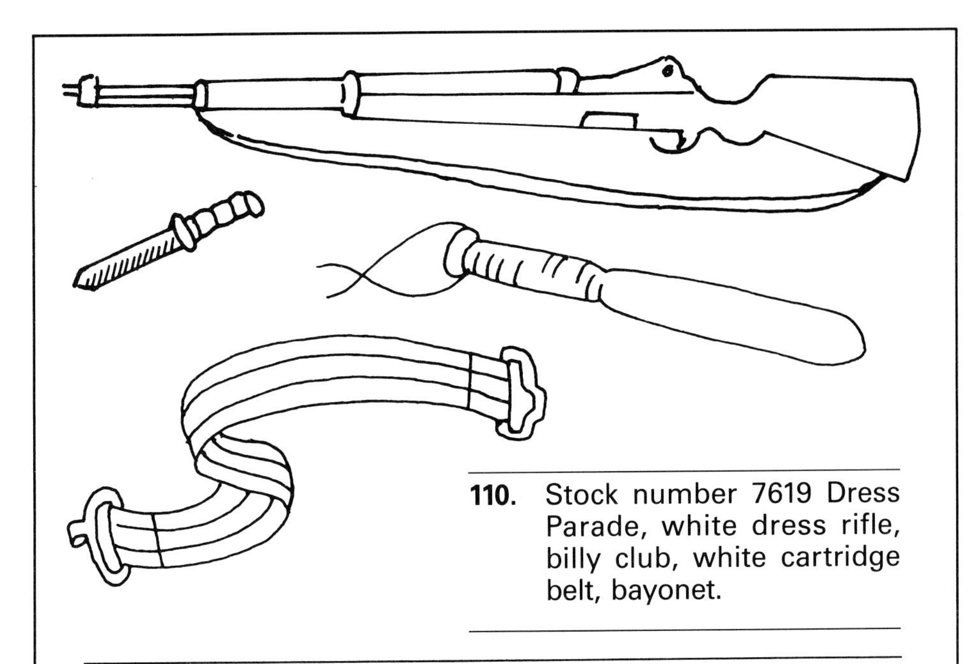

110. Stock number 7619 Dress Parade, white dress rifle, billy club, white cartridge belt, bayonet.

111. Stock number 7620 Deep Sea Diver, diving suit, double breast plate, diver's helmet, weighted belt, shoes, air pump, rubber hose, diver's tools, signal floats. *Lila Ayala Collection.*

112. Stock number 7621 Landing Signal Officer Set, safety-striped jumpsuit, cloth helmet with earphones, signal paddles, clipboard, binoculars, flare gun, goggles, pencil.

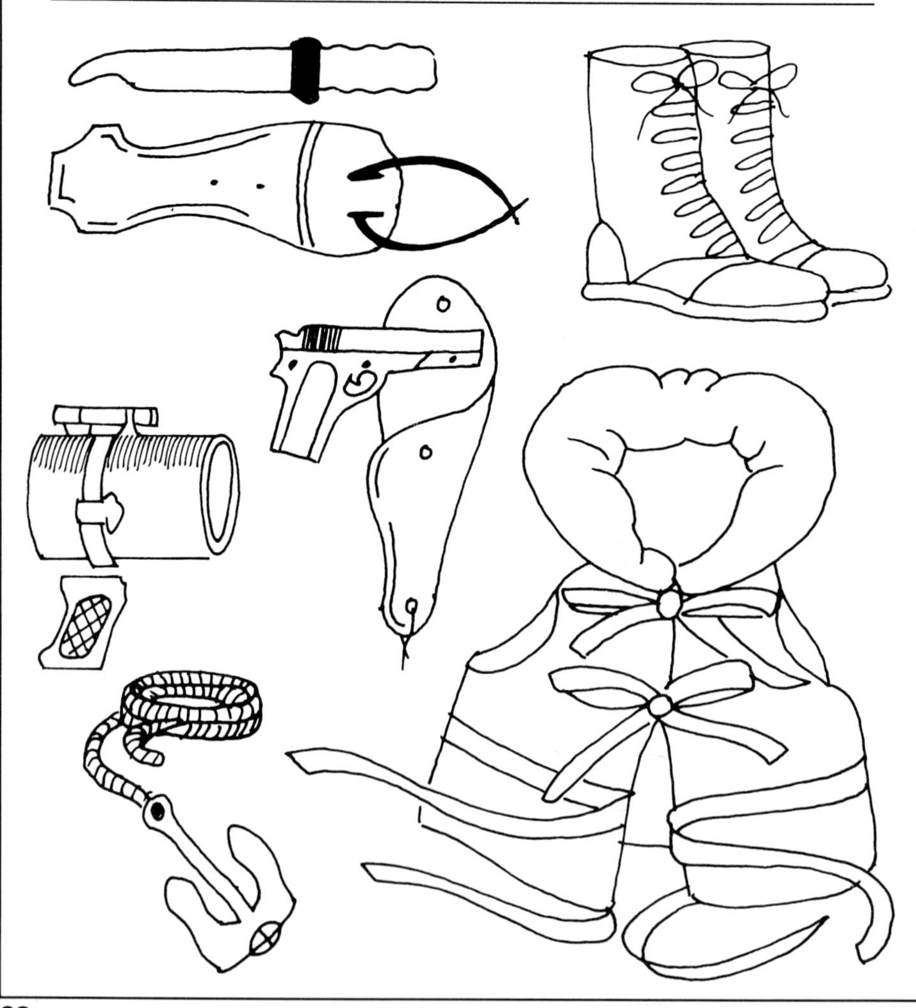

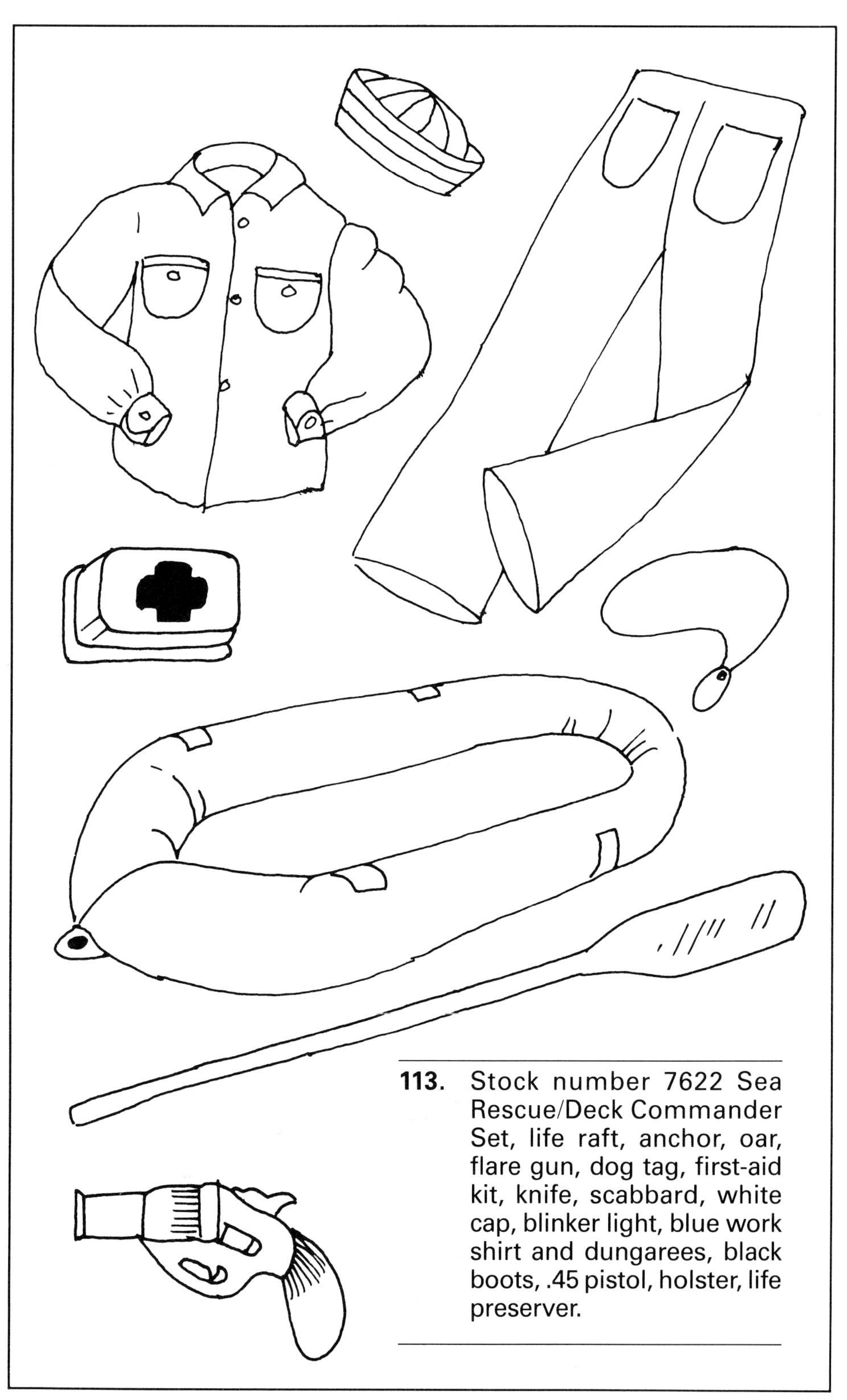

113. Stock number 7622 Sea Rescue/Deck Commander Set, life raft, anchor, oar, flare gun, dog tag, first-aid kit, knife, scabbard, white cap, blinker light, blue work shirt and dungarees, black boots, .45 pistol, holster, life preserver.

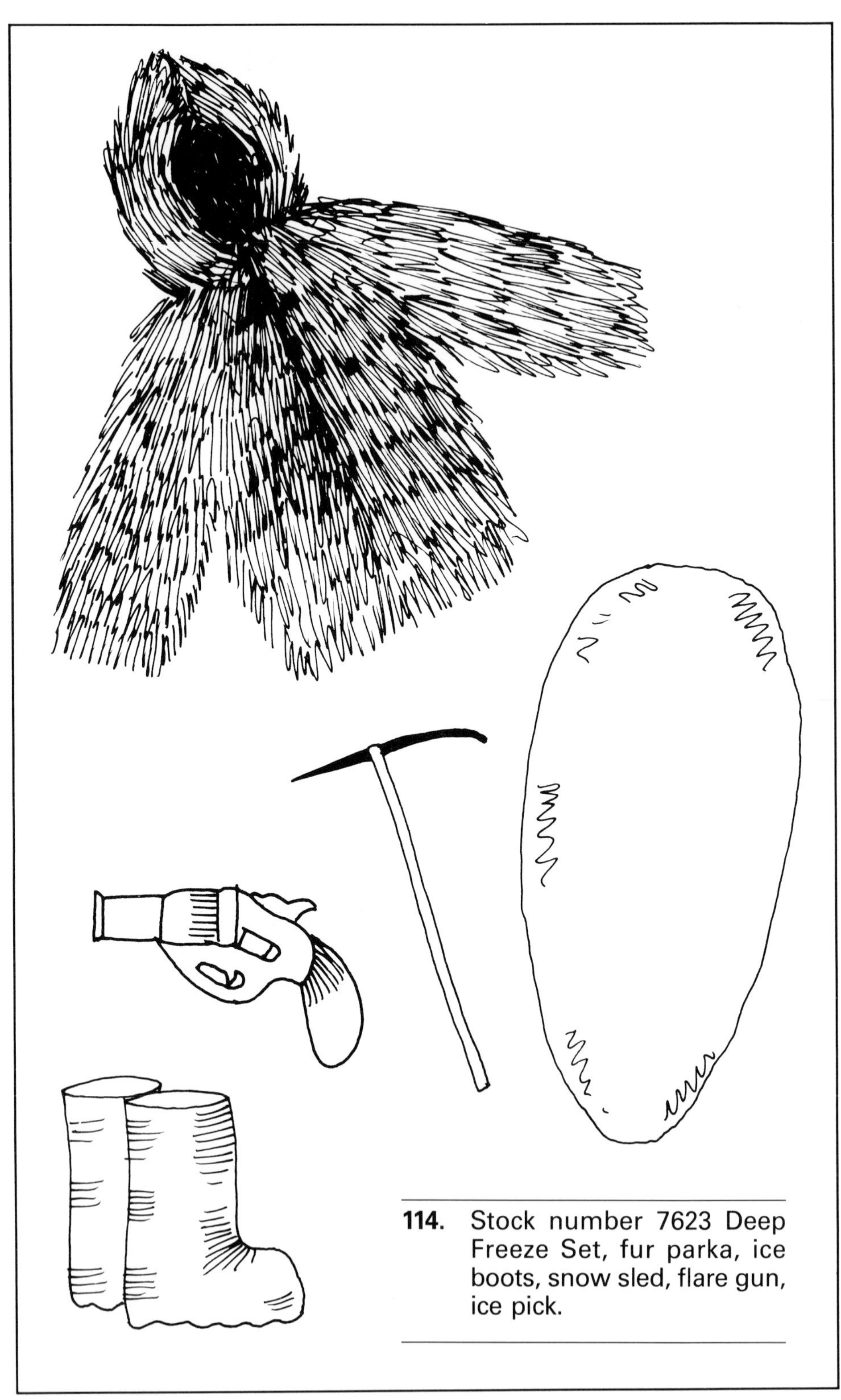

114. Stock number 7623 Deep Freeze Set, fur parka, ice boots, snow sled, flare gun, ice pick.

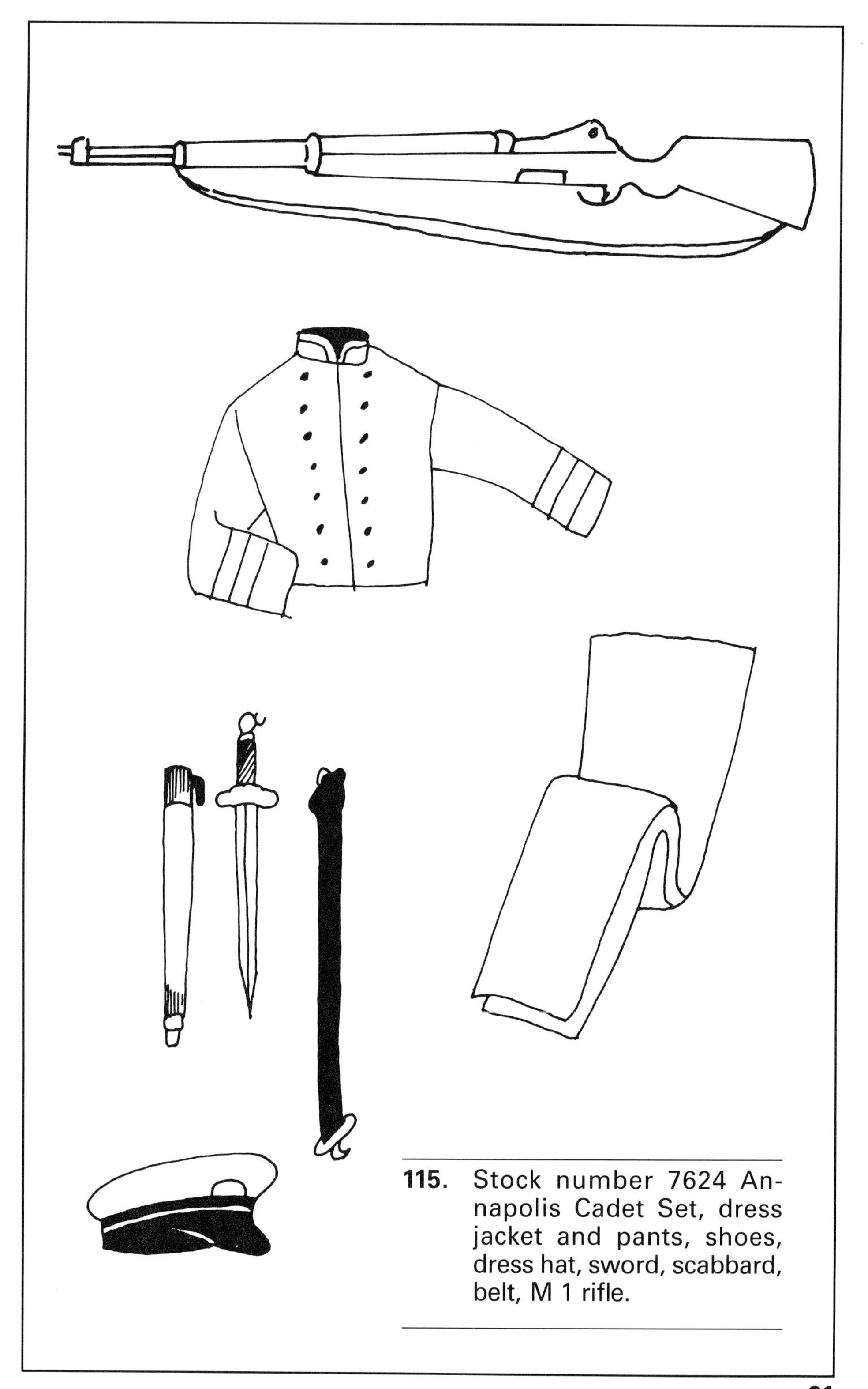

115. Stock number 7624 Annapolis Cadet Set, dress jacket and pants, shoes, dress hat, sword, scabbard, belt, M 1 rifle.

116. Stock number 7625 Breeches Buoy, breeches buoy, slicker jacket and pants, flare gun, blinker light.

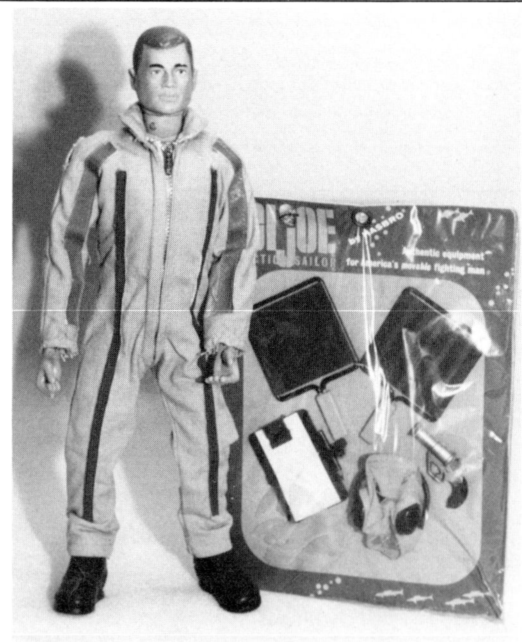

117. Stock number 7626 Landing Signal Officer Card, signal paddles, clipboard with pencil, flare gun, cloth helmet with earphones. *Lila Ayala Collection.*

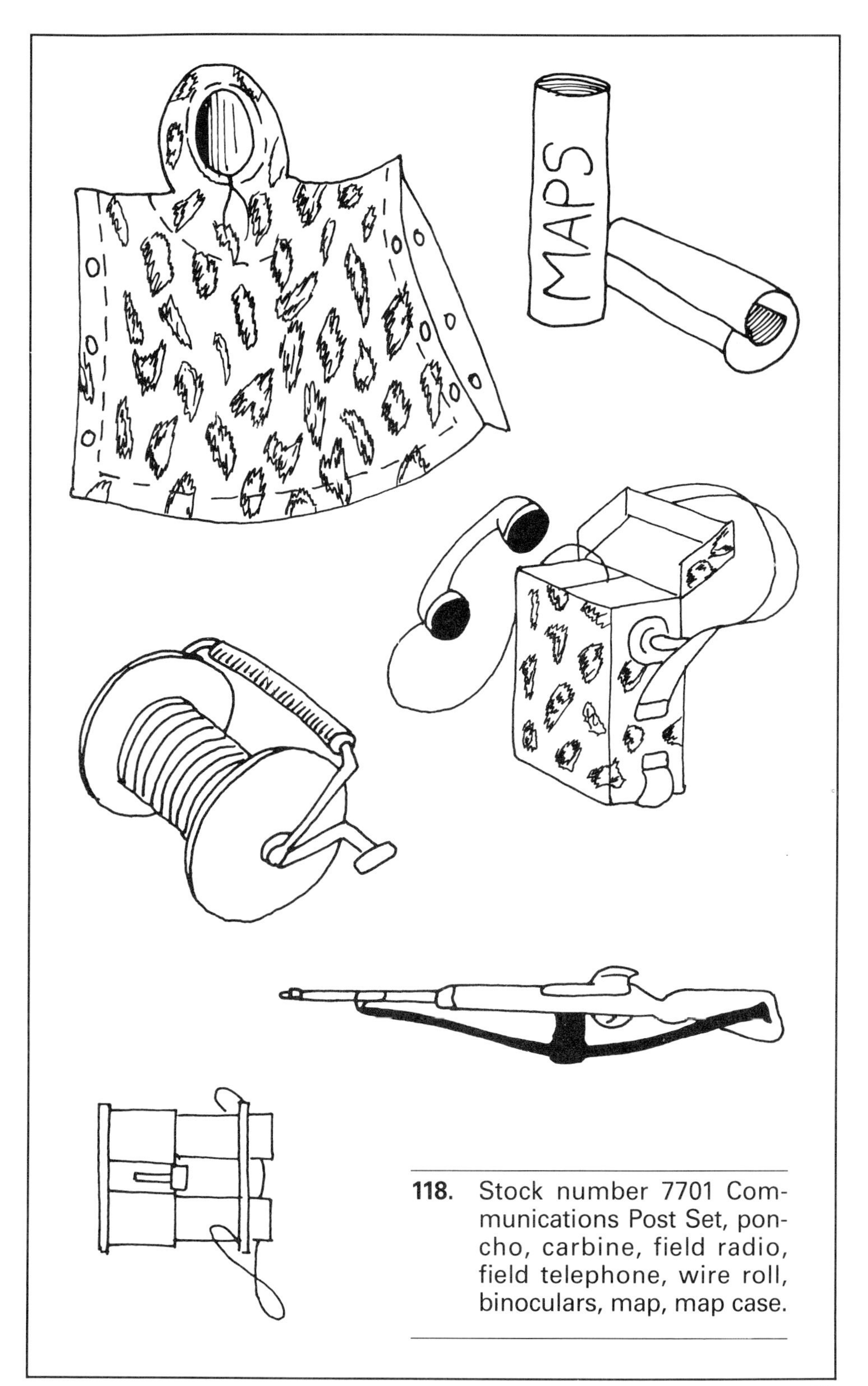

118. Stock number 7701 Communications Post Set, poncho, carbine, field radio, field telephone, wire roll, binoculars, map, map case.

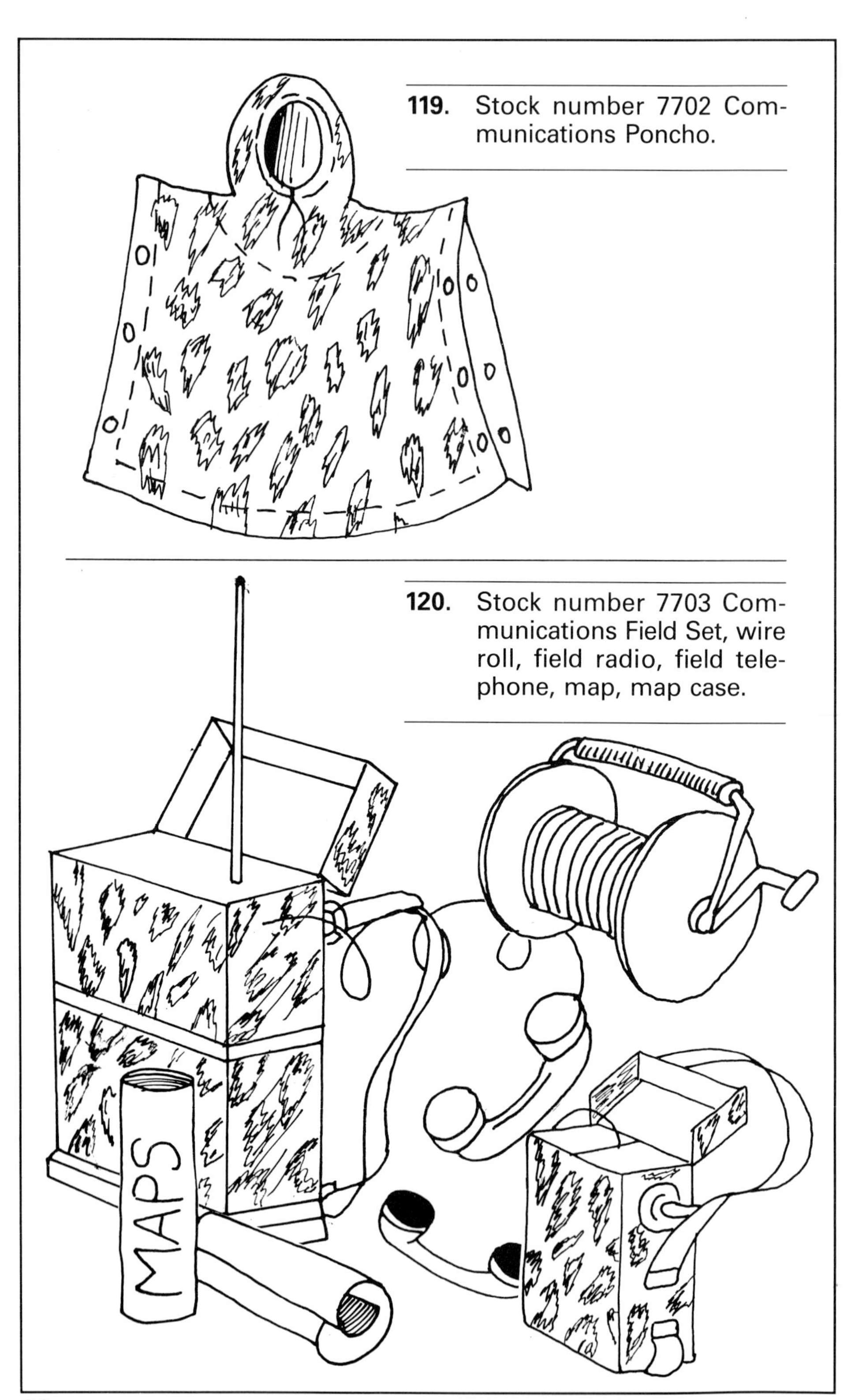

119. Stock number 7702 Communications Poncho.

120. Stock number 7703 Communications Field Set, wire roll, field radio, field telephone, map, map case.

121. Stock number 7704 Communications Flag Set, all service flags, Old Glory.

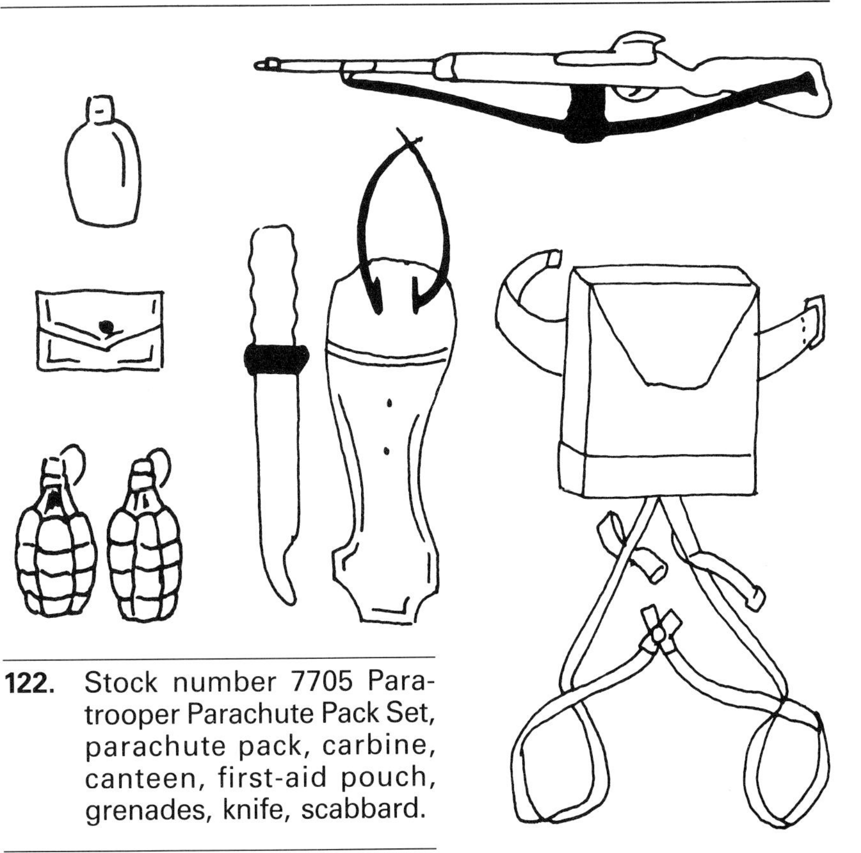

122. Stock number 7705 Paratrooper Parachute Pack Set, parachute pack, carbine, canteen, first-aid pouch, grenades, knife, scabbard.

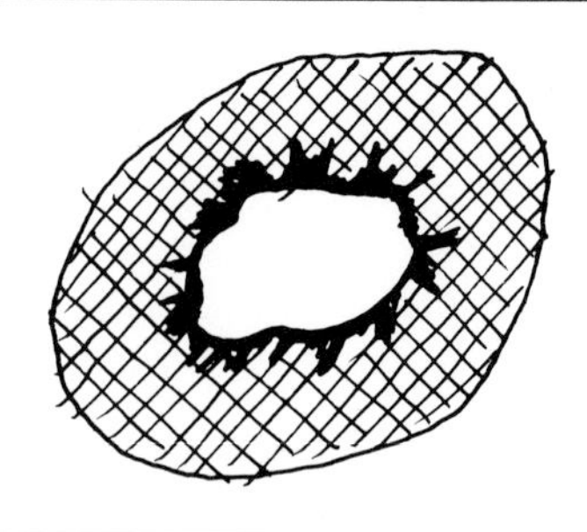

123. Stock number 7707 Paratrooper Helmet, helmet, camouflage netting, foliage.

124. Stock number 7708 Paratrooper Camouflaging Set, netting and foliage for tent.

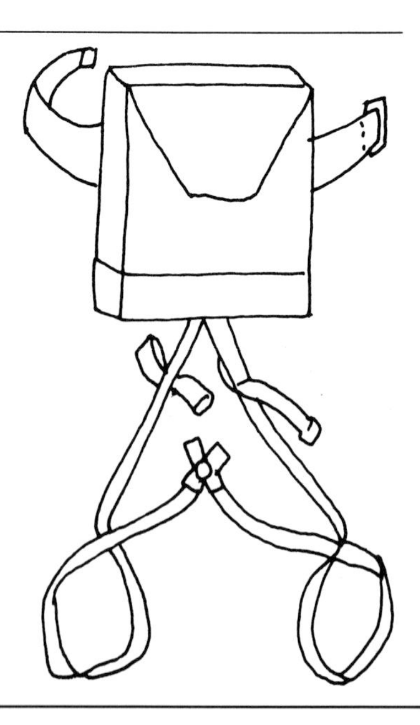

125. Stock number 7709 Paratrooper Parachute Pack.

126. Stock number 7710 Marine Dress Parade Set, dress jacket and pants with red stripe, white garrison cap, white pistol belt, M 1 rifle, black dress shoes, field manual.

127. Stock number 7710 Marine Dress Parade Set. *Lila Ayala Collection.*

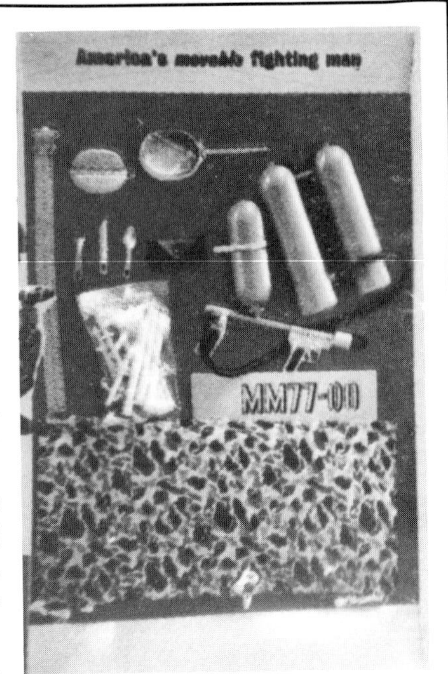

128. Stock number 7711 Beachhead Assault Tent Set, flamethrower, camouflage tent, pistol belt, first-aid pouch, mess kit, silverware. *Brochure - © 1965 Hassenfeld Bros., Inc.*

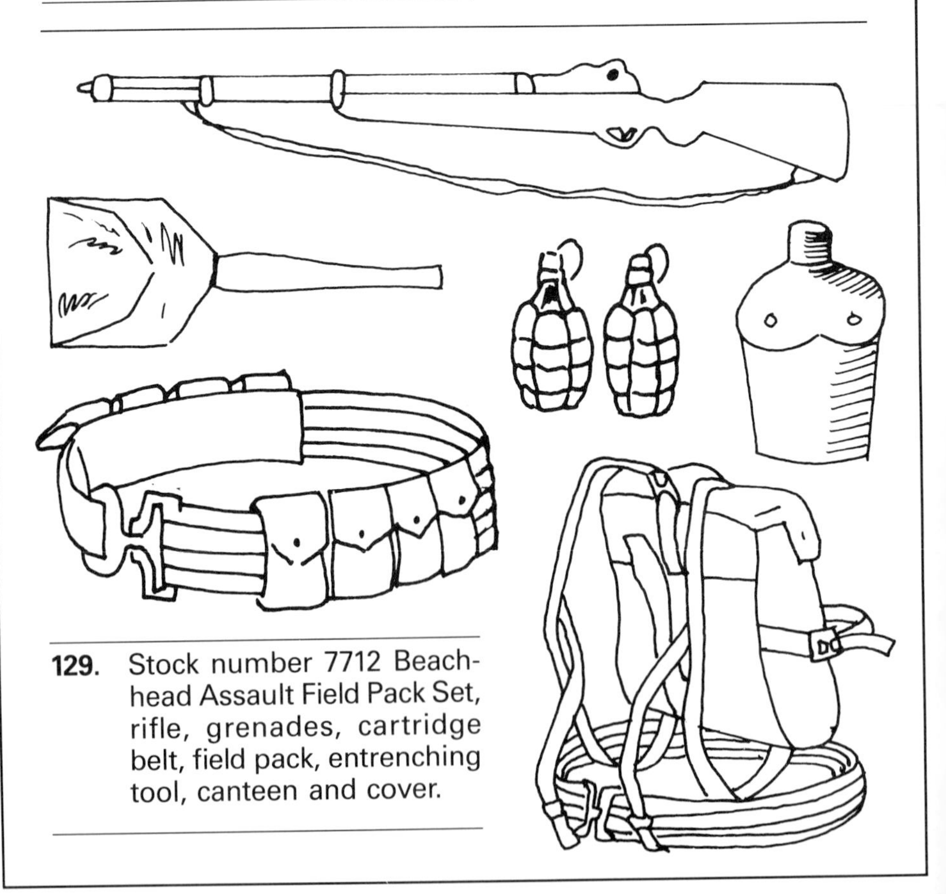

129. Stock number 7712 Beachhead Assault Field Pack Set, rifle, grenades, cartridge belt, field pack, entrenching tool, canteen and cover.

130. Stock number 7713 Beachhead Assault Field Pack.

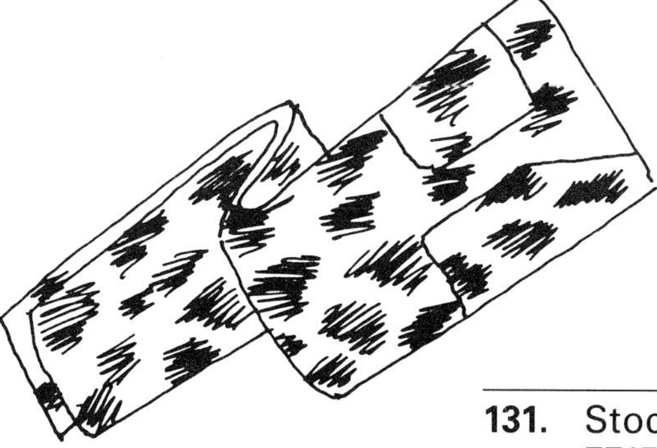

131. Stock numbers 7714 and 7715 Beachhead Assault Fatigue Shirt and Pants.

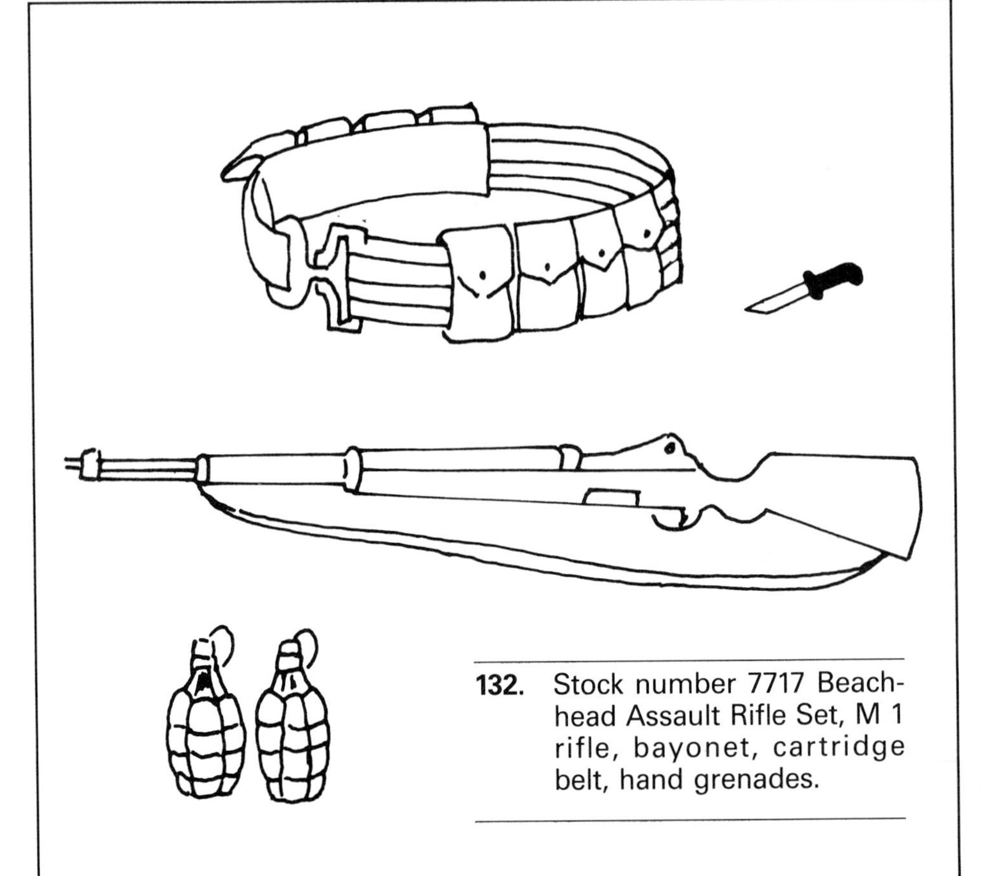

132. Stock number 7717 Beachhead Assault Rifle Set, M 1 rifle, bayonet, cartridge belt, hand grenades.

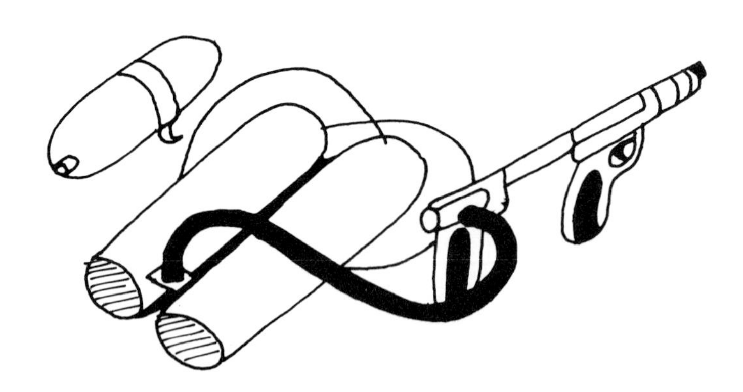

133. Stock number 7718 Beachhead Assault Flamethrower.

134. Stock number 7719 Marine Medic Set, stethoscope, splints, plasma bottle, bandages, Red Cross flag, medic bag, Red Cross armbands, crutch, stretcher.

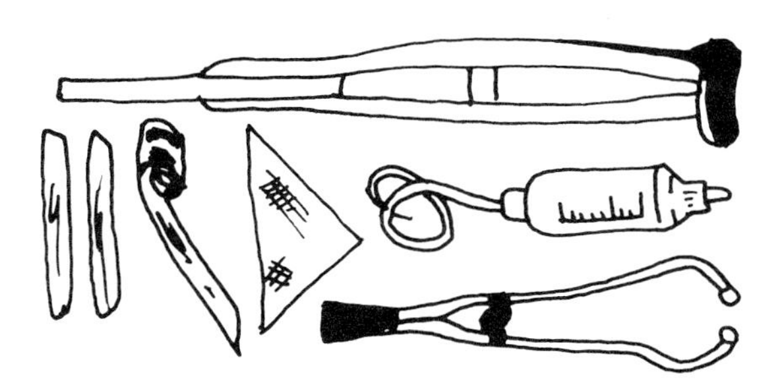

135. Stock number 7720 Medic Set, stethoscope, crutch, plasma bottle, bandage, splint package.

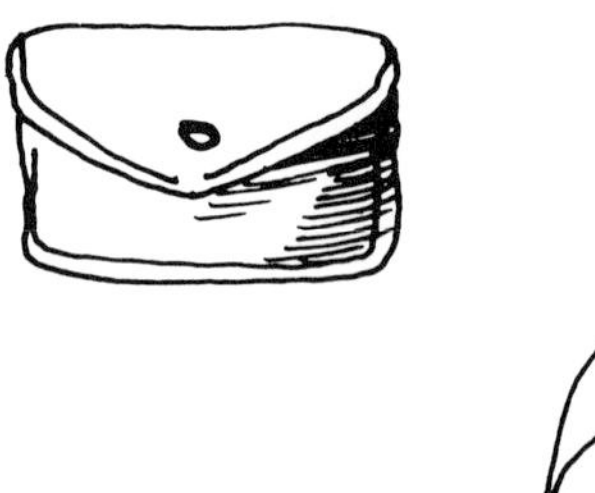

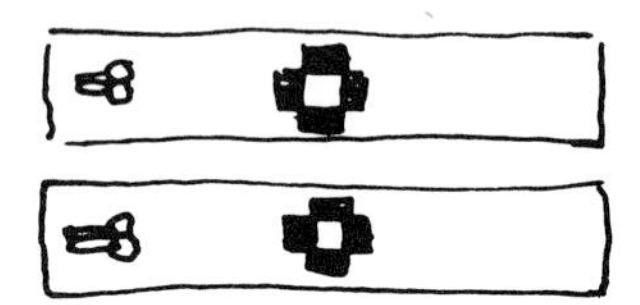

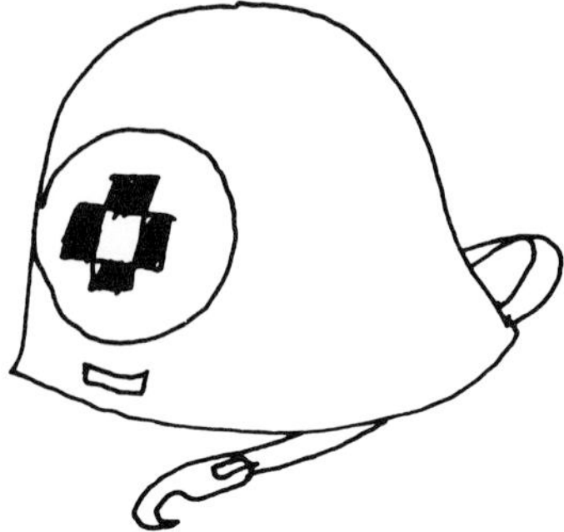

136. Stock number 7721 First-Aid Set, helmet with Red Cross decal, Red Cross armbands, first-aid kit.

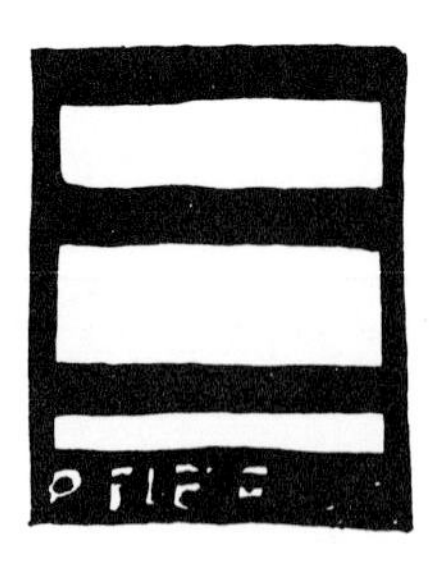

137. Stock number 7727 Weapons Rack.

138. Stock number 7730 Demolition Set, mine detector, power pack, earphones, mines. *Lila Ayala Collection.*

139. Stock number 7731 Tank Commander, brown leather jacket, tanker helmet, belt, machine gun and tripod, radio, helmet microphone, ammo box. *Lila Ayala Collection.*

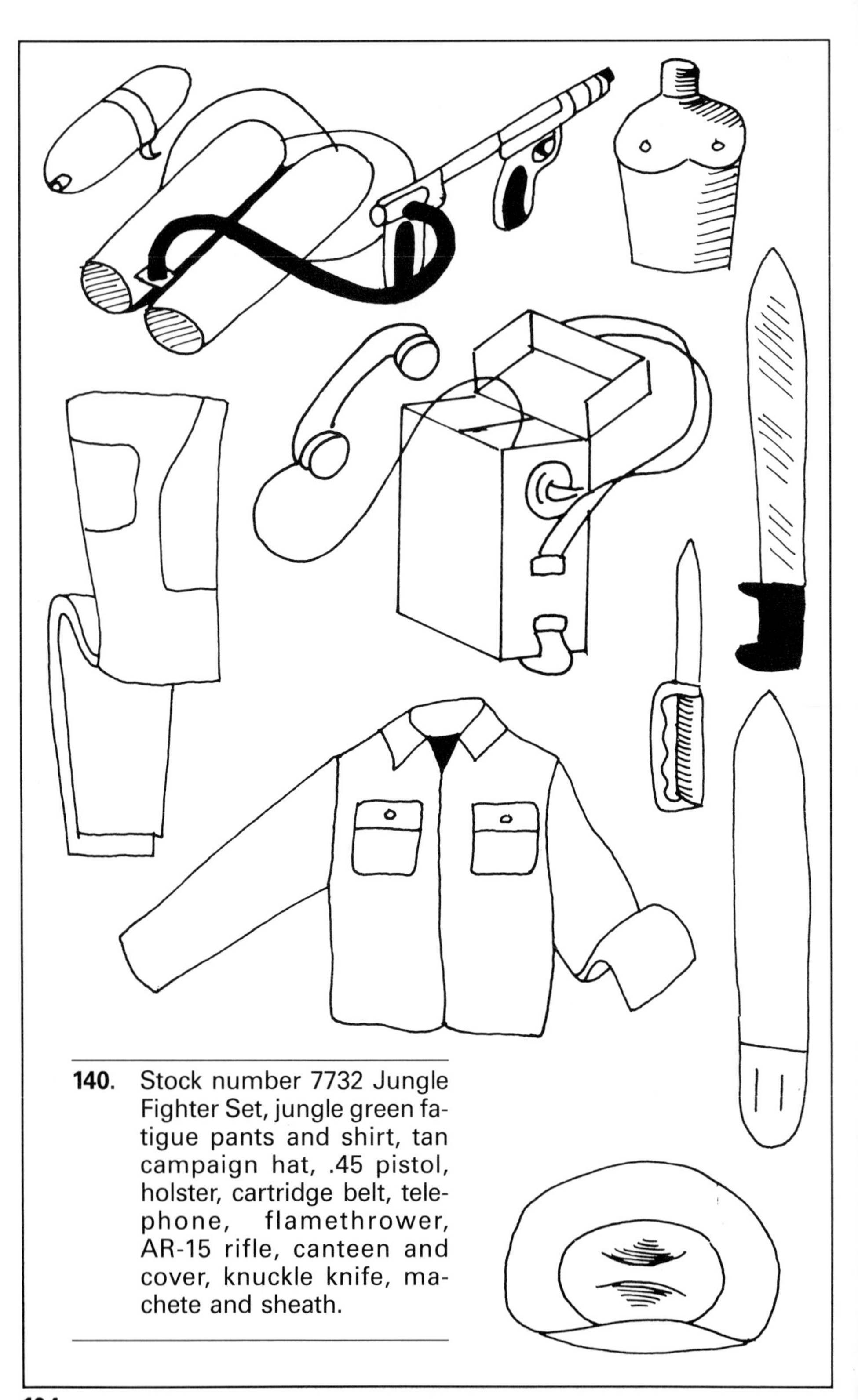

140. Stock number 7732 Jungle Fighter Set, jungle green fatigue pants and shirt, tan campaign hat, .45 pistol, holster, cartridge belt, telephone, flamethrower, AR-15 rifle, canteen and cover, knuckle knife, machete and sheath.

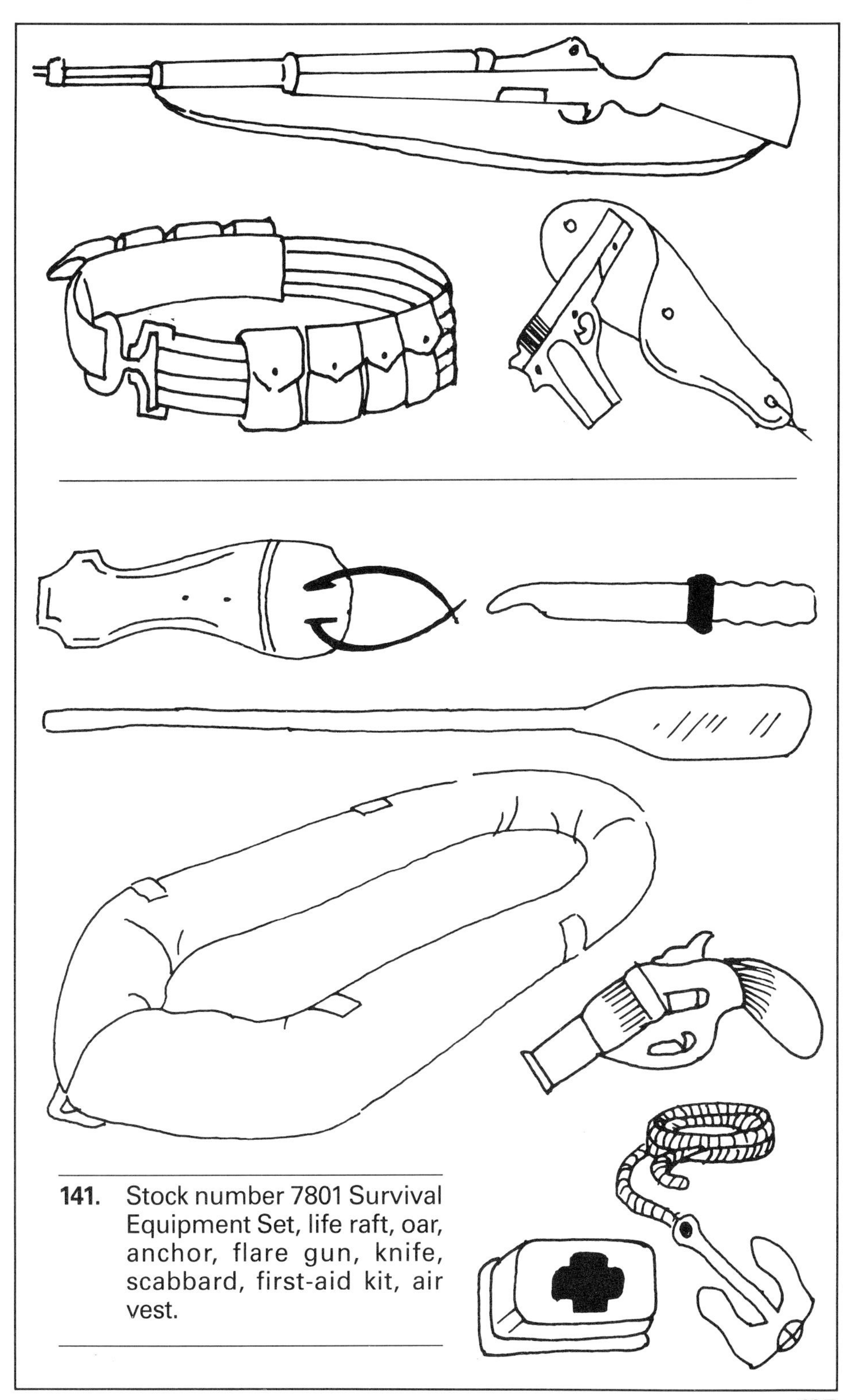

141. Stock number 7801 Survival Equipment Set, life raft, oar, anchor, flare gun, knife, scabbard, first-aid kit, air vest.

142. Stock number 7802 Survival Life Raft, life raft, anchor, oar.

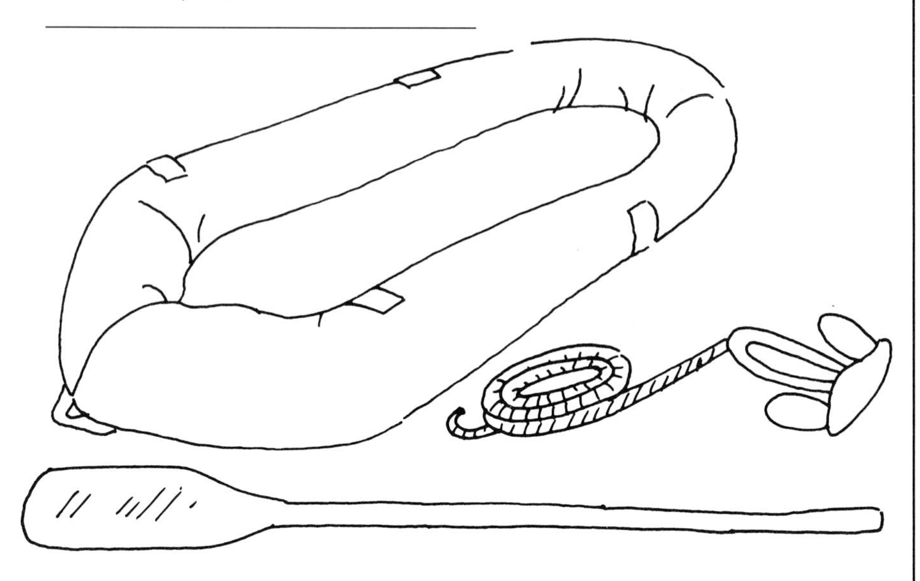

143. Stock number 7803 Air Force Dress Uniform, dress blue jacket, pants, shirt, tie, garrison hat, wings, captain's bars. *Lila Ayala Collection.*

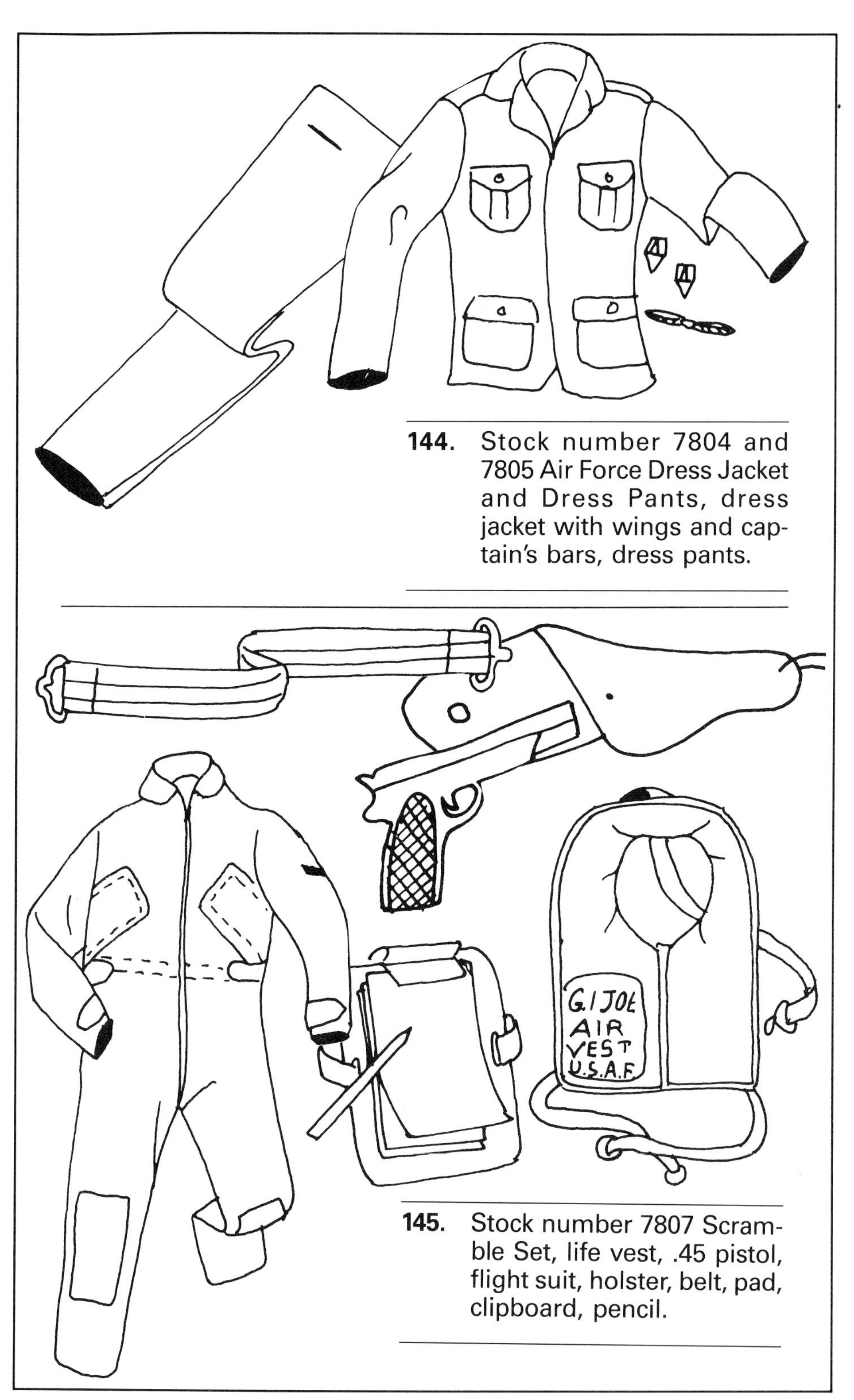

144. Stock number 7804 and 7805 Air Force Dress Jacket and Dress Pants, dress jacket with wings and captain's bars, dress pants.

145. Stock number 7807 Scramble Set, life vest, .45 pistol, flight suit, holster, belt, pad, clipboard, pencil.

146. Stock number 7808 Scramble Flight Suit.

147. Stock number 7809 Scramble Life Vest.

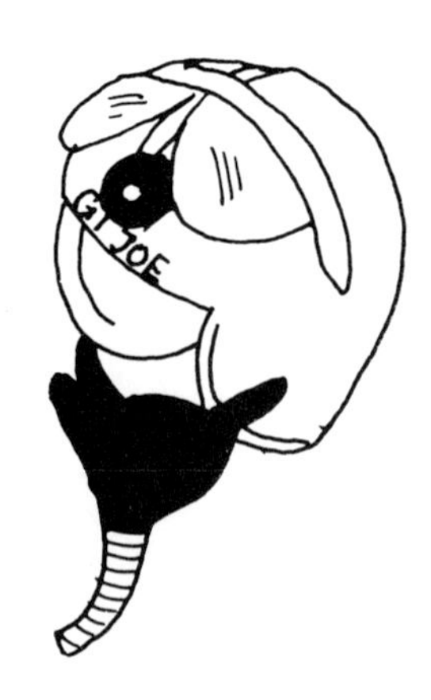

148. Stock number 7810 Scramble Helmet.

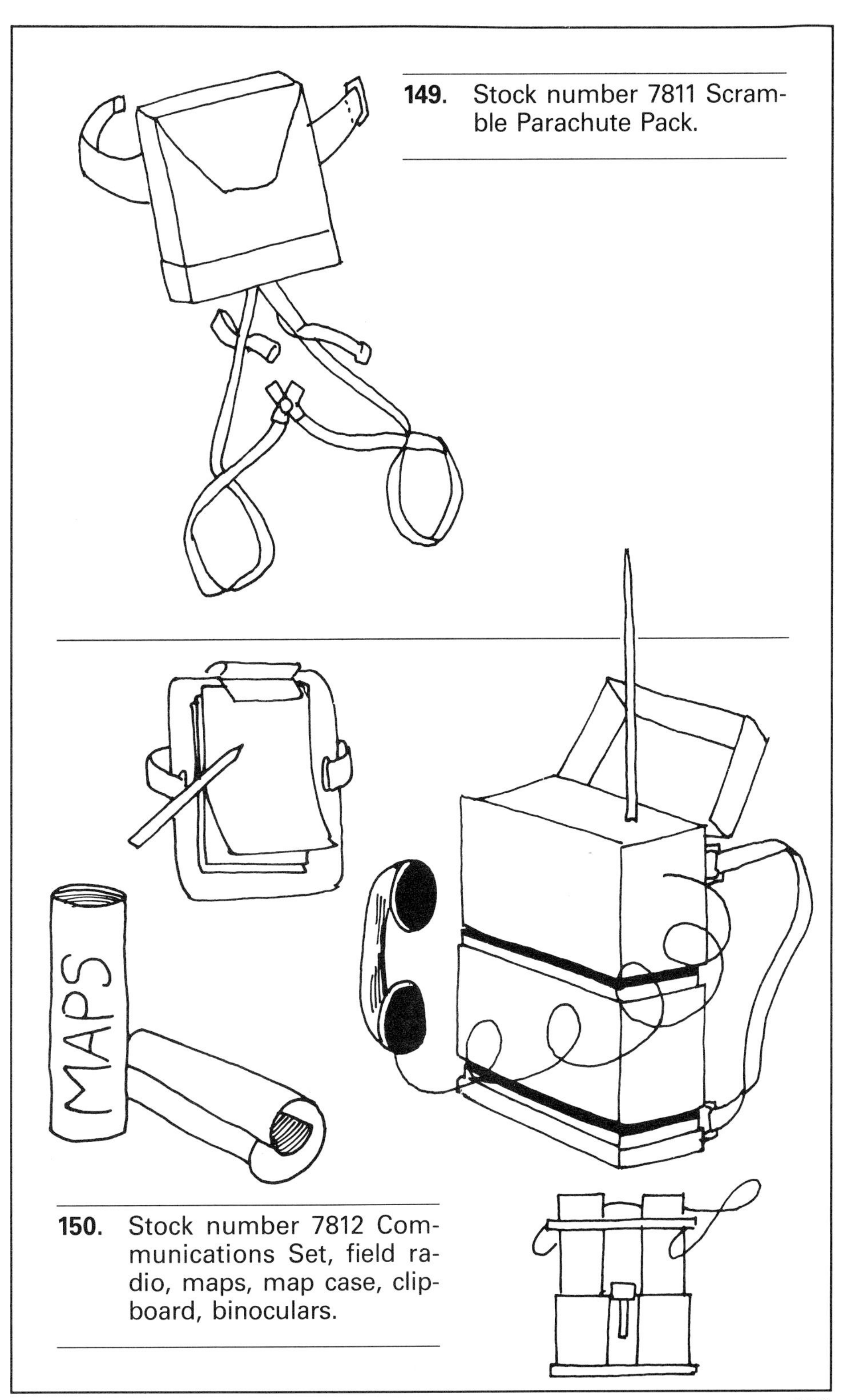

149. Stock number 7811 Scramble Parachute Pack.

150. Stock number 7812 Communications Set, field radio, maps, map case, clipboard, binoculars.

152. Stock number 7820 Fire Fighter/Crash Crew Set, metallic heat suit, gloves, protective hood, boots, belt with accessories, spray tank with nozzle. *Lila Ayala Collection.*

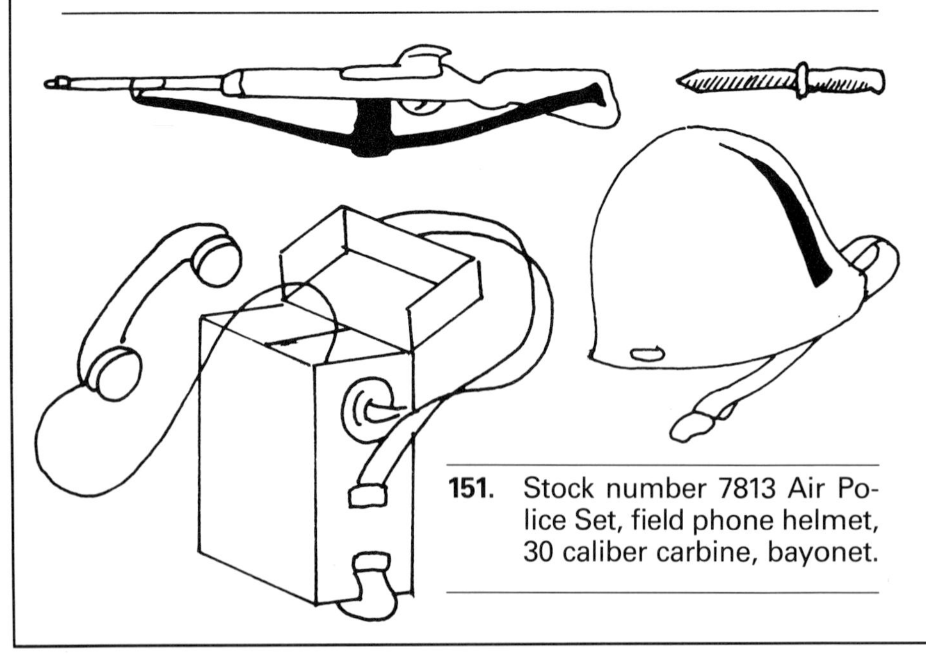

151. Stock number 7813 Air Police Set, field phone helmet, 30 caliber carbine, bayonet.

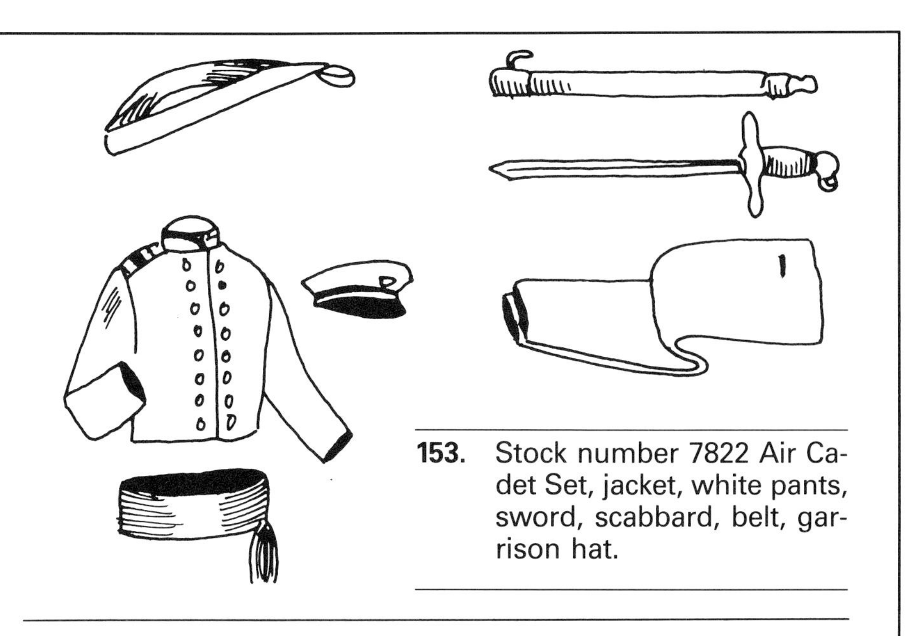

153. Stock number 7822 Air Cadet Set, jacket, white pants, sword, scabbard, belt, garrison hat.

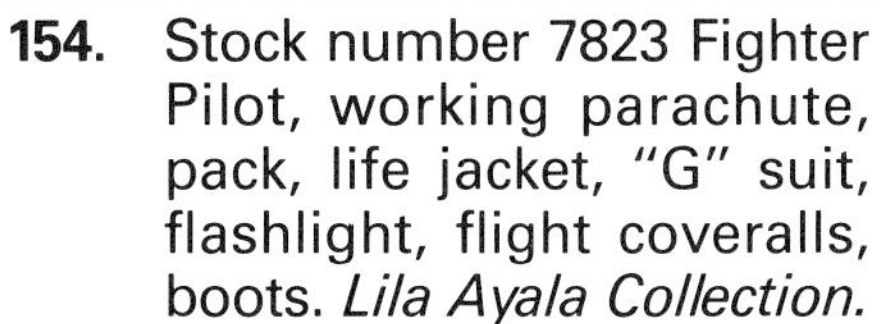

154. Stock number 7823 Fighter Pilot, working parachute, pack, life jacket, "G" suit, flashlight, flight coveralls, boots. *Lila Ayala Collection.*

155. Stock number 7824 Astronaut Suit, space suit, helmet, boots, gloves, space camera, hand propellent gun, chest pack, tether cord, air conditioner.

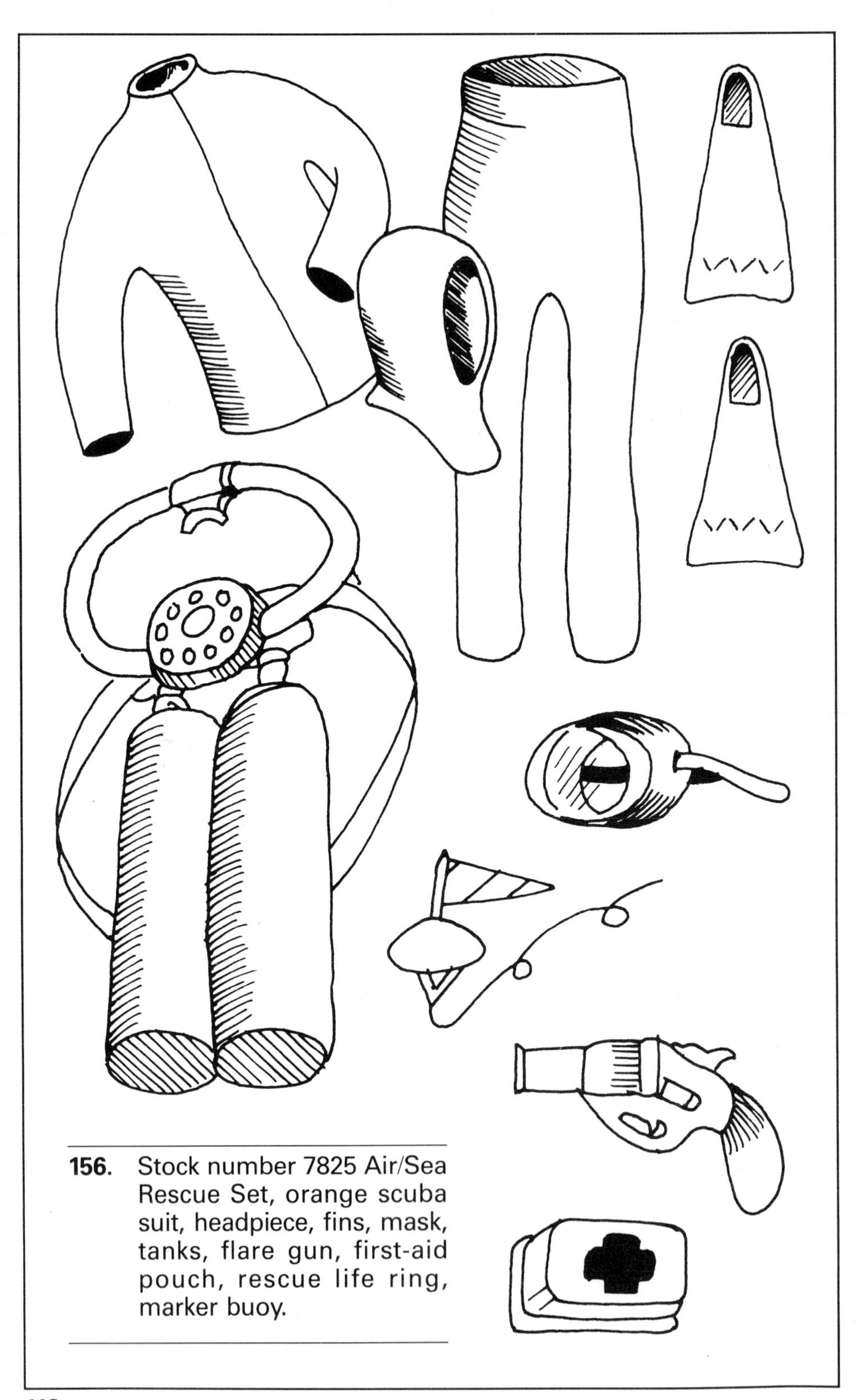

156. Stock number 7825 Air/Sea Rescue Set, orange scuba suit, headpiece, fins, mask, tanks, flare gun, first-aid pouch, rescue life ring, marker buoy.

VI. Extra Gear and Equipment

#7000 — Five Star Army Jeep with Trailer
#8000 — Foot Locker
#8020 — Astronaut Suit and Space Capsule Set
#8030 — Desert Patrol Jeep
#8040 — Crash Crew Fire Set
#8050 — Frogman Sea Sled Set
* Military Staff Car
* Motorcycle
* Side Car/Cycle
* Personnel Carrier/Mine Sweeper
* Navy Jet
* Carry Case/Sentry Post (Sears Exclusive)

* — Stock numbers unknown

157. Stock number 7000 Five Star Army Jeep with Trailer from brochure ©1965 Hassenfeld Bros., Inc.

158. Stock number 8000 Foot Locker from brochure ©1965 Hassenfeld Bros., Inc.

159. Stock number 8020 Astronaut Suit and Space Capsule Set, floating space capsule featuring sliding canopy, control panel, retro pack and communications plug-in, authentic space suit, astronaut helmet, space gloves and boots, life raft, oar, flotation collar.

160. Stock number 8020. *Lila Ayala Collection.*

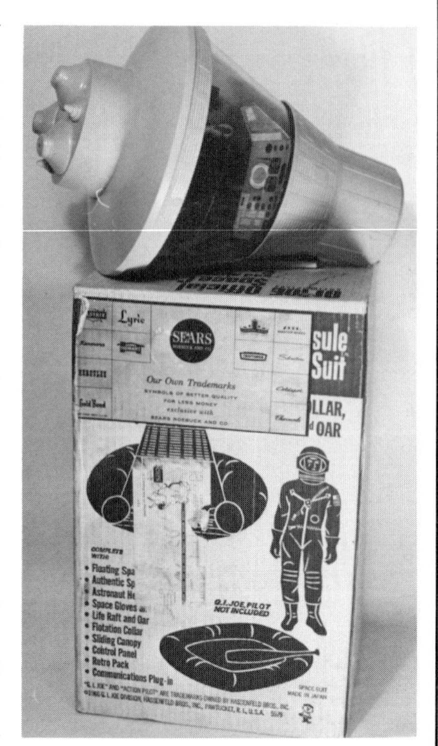

VII. Adventure Sets and Outfits

#7302 — Copter Rescue
#7303 — Secret Rendezvous
#7308 — Jungle Ordeal
#7309 — Undercover Agent
Secret Mission
Dangerous Climb
Desert Explorer
] Pack assortment, same stock numbers

#7319-2 — Magnetic Flaw Detector
#7319-4 — Thermal Terrain Scanner
#7319-6 — Seismograph
#7340 — Missile Recovery
#7342 — High Voltage Escape
#7350 — Rescue Raft
#7351 — Fire Fighter
#7352 — Lifeline Catapult
#7360 — Escape Cart
#7361 — Flying Rescue
#7362 — Signal Flasher
#7363 — Turbo Copter
#7364 — Drag Bike
#7370 — Demolition/Dangerous Removal
#7371 — Smoke Jumper
#7372 — Karaté
#7373 — Jungle Survival
#7374 — Emergency Rescue
#7375 — Secret Agent
#7411 — Secret Spy Mission to Spy Island
#7412 — Danger of the Depths
#7413 — Revenge of the Spy Shark
#7414 — Black Widow Rendezvous
#7420 — Attack at Vulture Falls
#7421 — Jaws of Death
#7422 — Eight Ropes of Danger
#7423 — Fantastic Freefall
#7436 — White Tiger Hunt
#7437 — Capture of the Pigmy Gorilla
#7439 — Devil of the Deep
#7440 — Sky Dive to Danger

#7441 — Secret of the Mummy's Tomb
#7450 — Talking Fate of the Troubleshooter
#7480 — Sea Wolf Submarine
#7495 — G.I. Joe Training Center
#8040 — Atomic Man Secret Mountain Outpost
#8208 — Secret Courier

All stock numbers of the Adventure Sets and Outfits listed below are unknown.

Hurricane Spotter
Raging River Dam-up
Spacewalk Mystery Adventure
Fight for Survival
Scout Helicopter
Adventure Team Headquarters
All-Terrain Vehicle
Mobile Support Unit
Recovery of the Lost Mummy
Jungle Kit
System Control Headquarters
Mobile Support Vehicle
Training Center
Tank
Capture Copter Adventure
Tote-a-Tent
Mobile Sentry Tower
Helicopter
Signal ATV
Action Sea Sled
Giant Helicopter
Rugged Amphicat
Sporty Cycle
Search for the Abominable Snowman
Mystery of the Boiling Lagoon
Escape from Danger
Danger Alert
Search for the Stolen Idol
Secret Stronghold
Sky Hawk
Shark's Surprise
Skin Diver Rescue
Safari Into Doom
Fire Fighter Rescue
Sandstorm Survival
Aerial Recon
Volcano Jumper
Radiation Detection

161. Stock number 7309 Desert Explorer, 7302 Copter Rescue, 7309 Dangerous Climb, 7303 Secret Rendezvous.

162. Stock number 7309 Secret Mission, 7309 Undercover Agent, 7309 Dangerous Climb.

163. Stock number 7319-4 Thermal Terrain Scanner, 7319-2 Magnetic Flaw Detector, 7319-6 Seismograph.

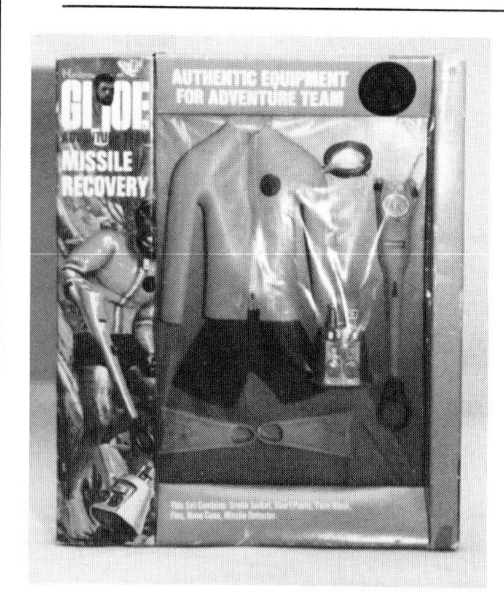

164. Stock number 7340 Missile Recovery.

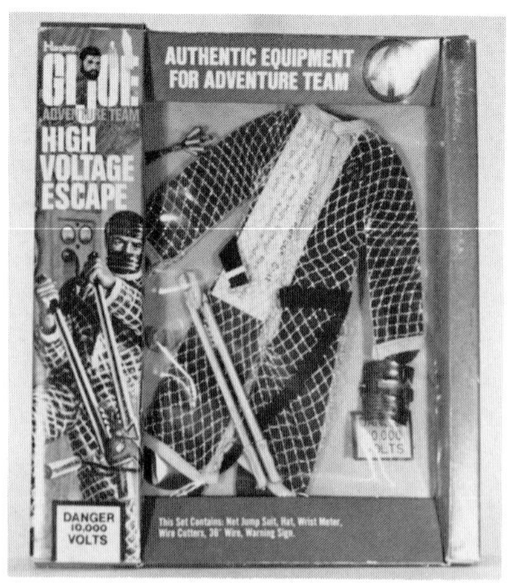

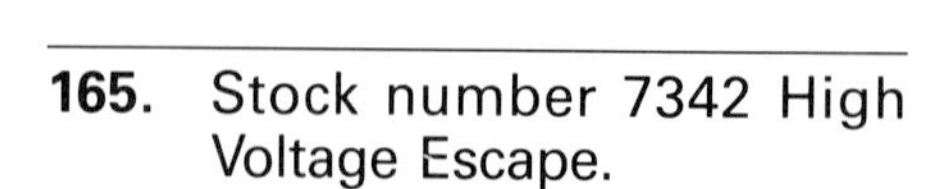

165. Stock number 7342 High Voltage Escape.

166. Stock number 7370 Demolition/Dangerous Removal.

167. Stock number 7371 Smoke Jumper.

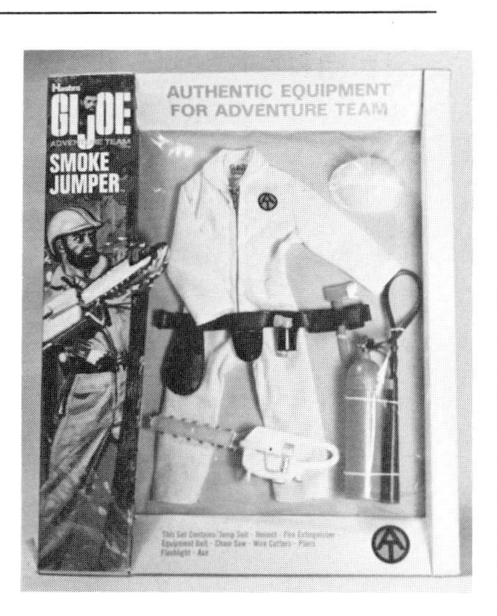

168. Stock number 7372 Karaté.

169. Stock number 7373 Jungle Survival.

170. Stock number 7374 Emergency Rescue.

171. Stock number 7375 Secret Agent.

172. Stock number 7421 Jaws of Death.

173. Stock number 7422 Eight Ropes of Danger.

174. Stock number 7423 Fantastic Freefall.

175. Stock number 7436 White Tiger Hunt.

176. Stock number 8040 Atomic-man Secret Mountain Outpost.

177. Stock number unknown. Canadian Mountie Uniform. *Lila Ayala Collection.*

178. Stock number unknown. Recovery of the Lost Mummy. *Sears 1972 Christmas catalog.*

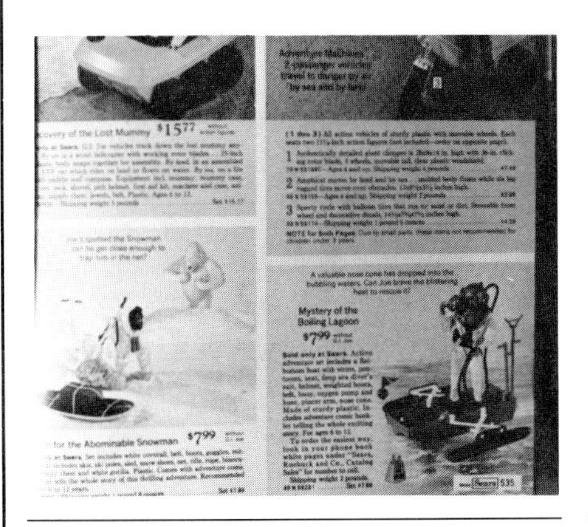

179. Stock numbers unknown. Search for the Abominable Snowman and Mystery of the Boiling Lagoon. *Sears 1974 Christmas catalog.*

180. Stock numbers unknown. G.I. Joe Mobile Support Unit, Giant Helicopter and Amphicat. *Sears 1972 Christmas catalog.*

VIII. Miscellaneous

Phrases — each talking figure:

Army:

1. G.I. Joe, U.S. Army, reporting for duty.
2. Take the Jeep and get some ammo, fast.
3. Cover me, I'll get that machine gun.
4. Medic, get that stretcher up here.
5. Enemy planes. Hit the dirt.
6. This is Charlie Company, send reinforcements.
7. Take Hill 79, move out.
8. All units commence firing.

Air Force:

1. G.I. Joe, U.S. Air Force reporting for duty.
2. Crash crew, emergency — runway four.
3. Sirens, paramedics scramble.
4. All systems go. Ten seconds to re-entry.
5. Over target, bombs-away.
6. Enemy fighters, 12 o'clock high.
7. Runway clear, prepare for take-off.
8. Engine two on fire, bail out.

Marine:

1. G.I. Joe U.S. Marines, reporting for duty.
2. We must hold this position, dig in.
3. Paratroopers hook up, Geronimo.
4. Man the machine guns.
5. Marines hit the beach.
6. Enemy patrols sighted.
7. Let's go Leathernecks, move out.
8. Prepare wounded for helicopter pick-up.

Sailor:

1. G.I. Joe, U.S. Navy reporting for duty.
2. Damage control, fire in the engine room.
3. Enemy submarines off port bow.
4. Now hear this, prepare to launch aircraft.
5. (whistle) Salvage divers report to the bridge.
6. Capsule sighted, prepare to pick up astronauts.
7. Frogmen, check your scuba gear.
8. (three buzzers) Man your battle stations.

G.I. Joe Commander:

1. I've got a tough assignment for you.
2. The Adventure Team has the situation controlled.
3. The Adventure Team is needed in Africa.
4. We must get there before dark, follow me.
5. Mission accomplished, good work, men.
6. Set up Team headquarters here.
7. Contact Adventure Team Headquarters, right away.
8. This is going to be rough, can you handle it?

Astronaut:

1. Ten seconds to lift-off and counting.
2. We have ignition, lift off.
3. Confirm second-stage ignition, entering Lunar orbit.
4. Prepare lunar Module.
5. Main chute is open.
6. Landing party is now on Moon.
7. Splashdown on target.
8. Contain Lunar orbit.

IX. Super Joe — Body marks, stock numbers

©1977 Hasbro®
Pat. Pend (back of swim trunks)
#7511 — Edge of Adventure, with yellow hooded jacket, rope, black pants, pickax.
#7512 —
#7513 — Paths of danger, with blue and red jumpsuit, yellow spiral patch on chest, binoculars.
#7514 — Emergency Rescue, with silver apron, pinchers, "glow" rock.

181. Stock numbers 1965 The Secret Mission to Spy Island, 1966 Secret of the Mummy's Tomb, 1967 The Search for the Stolen Idol, 1968 The Rescue from Adventure Team Headquarters. Adventurers on 45 rpm records — Book and Record by Peter Pan Records.

182. Stock number 1965 Book and Record Adventures on 33⅓ rpm record by Peter Pan Records. Three adventures on one record. *Lila Ayala Collection.*

183. Left to right: Stock numbers unknown, stock number 7514 Emergency Rescue, Standard Uniform, 7511 Edge of Adventure, stock number unknown. *Lila Ayala Collection.*

Index